James Ward's blog, *I Like Boring Things*, has featured in the *Independent* and the *Observer* and on the BBC website. He is co-founder of the Stationery Club in London. His annual Boring Conference has featured in the
Wall Street Journal and the *Sun*
Adventures in Stationery is his fir

Adventures in STATIONERY

A Journey Through Your Pencil Case

P

PROFILE BOOKS

This paperback edition published in 2015

First published in Great Britain in 2014 by
Profile Books Ltd
3 Holford Yard
Bevin Way
London WC1X 9HD
www.profilebooks.com

Illustrations on pages 27, 42, 45, 61, 71, 75, 91, 99, 101, 108, 123, 147, 183, 192, 210, 217, 219, 239, 257 and 263 reproduced courtesy of iStock. All other text illustrations from the author's collection and other copyright-free sources. Additional photography of items from the author's collection by Micheline Mannion at Profile Books.

A CIP catalogue record for this book is available from the British Library.

ISBN 978 1 84668 616 0
eISBN 978 1 84765 871 5

Text design by *sue@lambledesign.demon.co.uk*
Typeset in Iowan by MacGuru Ltd *info@macguru.org.uk*

Printed by CPI Group (UK) Ltd, Croydon CR0 4YY

CONTENTS

ACKNOWLEDGEMENTS

This being my first book, I was surprised and slightly bewildered by the number of different stages involved in turning a bundle of words into a book, and it feels slightly unfair that there is only one name on the front cover when so many other people were involved in the process.

There wouldn't be a book at all were it not for Andrew Gordon at David Higham Associates, and I am massively grateful to him for his support and encouragement. Thanks also to Marigold Atkey.

Enormous thanks to everyone at Profile Books (past and present) who was involved in turning my scraps of paper into a book, including Lisa Owens, Rebecca Grey, Daniel Crewe, Anna-Marie Fitzgerald, Paul Forty and Andrew Franklin. Particular thanks must go to Sarah Hull for her incredible job of editing my text and shaping it into something coherent. Thank you also to copy-editor Fiona Screen for correcting my many mistakes.

Given that the subject matter of this book is all about the pleasure of physical objects, it was important that the design and layout of the book were right and so thanks are due to Pete Dyer at Profile Books for the cover and – with Micheline Mannion – all his work on compiling the images throughout the book.

A book like this would be impossible without the assistance of the many brands and stationery companies featured in its pages. Particular thanks go to STABILO, Bostik, BIC, Helix, 3M, Ryman, Sheaffer and Henkel for supplying company information. I'd also like to thank Kevan Atteberry, Geoff Nicolson and Spence Silver for answering my questions.

I also need to thank Neal at Present & Correct for running the most wonderful stationery shop in London. Go there. Go there now.

Thanks also to Bob Patel and everyone at Fowlers Stationers in Worcester Park for inspiring a life-long passion.

If Ed Ross hadn't created the #stationeryclub hashtag on Twitter, it's unlikely this book would exist. Thank you Ed.

Finally, I'd like to thank Natassia Caffery for her patience during the weekends and evenings when I was lost in research and surrounded by bits of paper and Post-it Notes.

(Oh, I also need to thank my mum for telling everyone that she met in the last year or so to buy this book.)

CHAPTER 1

Velos 1377 revolving desk tidy

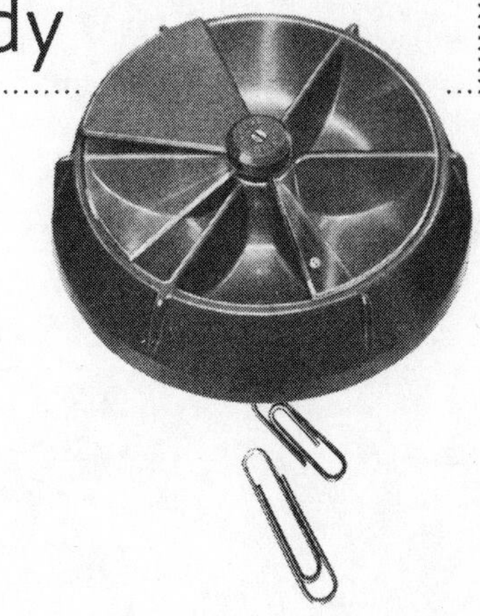

I grew up in Worcester Park, a small town in Surrey. As a child, I would regularly visit Fowlers, an independent stationer on the high street. This shop had always interested me. Yes, there was a bigger WHSmith at the bottom of the hill, and yes, I spent quite a lot of time looking at pens in there too, but it wasn't the same. Fowlers seemed more serious about stationery. WHSmith had books and magazines and toys and sweets and videos. Fowlers were more dedicated. They sold different types of clips and tags, not the sort they sold in Smiths. They had foolscap suspension files. Office supplies. Grown-up things. It was a quiet shop. Ponderous. A bit like a library. Or at least half of it was. The other half was given over to greetings cards and wrapping paper and cheap gifts. That side didn't interest me. But this half – my half – captivated me with its racks of pens and pencils. I would spend long periods of time here studying these objects. Picking them up, turning them over in my hand. Sometimes I'd even buy something.

A few years ago, I returned to Fowlers. It was still the same as I remembered it; very little had changed. Even the man behind the counter was the same. There wasn't anything in particular that I needed, but I wandered around the store, letting my eyes drift from item to item. Behind some packets of record cards (Silvine, 204 mm × 127 mm, ruled), I saw a rather tatty-looking box. It was square, about six inches by six inches and about two inches tall. On the top, in white lettering on a lurid pink background, it said 'VELOS 1377 – REVOLVING DESK TIDY', and underneath in slightly smaller writing 'Six compartments with cover' next to a black-and-white picture of the revolving desk tidy itself. I picked it up. I'd never heard of Velos before, and looking at the box, I'm not surprised. This desk tidy was quite possibly older than me. The box looked like it must have been from the late 1970s. It was covered in dust. It didn't look like anyone had picked it up for years; it had just been stuck at the back of a shelf, forgotten about. I had to own it. I took it to the counter to pay. The man behind the counter looked for a barcode, but there wasn't one – it came from a time before barcode scanners. Fortunately, it had a faded price sticker in one corner: £5.10 (this couldn't be the original price, surely? It was too expensive. When had the price been changed?). The man behind the counter shrugged, keyed the price into the till and, as I paid, he made a note of the item in a little stock book.

When I got home, I opened the box carefully – I didn't want to tear it. Inside, there it was: the 1377 Revolving Desk Tidy. The desk tidy was in perfect condition – not surprising as, despite its age, I'd effectively bought it as new. Small and round and 'moulded in high impact styrene', it had a transparent cover showing its six compartments. The round 'tidy' was divided into six segments 'for all types of small sundries', and looked a bit like a grapefruit cut across the middle. The cover had an opening the same size as one of the compartments and a little lid you could slide across to open or close. You could spin the whole thing round so whichever compartment you wanted to access was under the opening, allowing you to reach in and take some paperclips or drawing pins or whatever else you decided to fill your six compartments with (the picture on the cover showed the desk tidy empty; there was no 'serving suggestion' – Velos customers were trusted to use their initiative).

I filled my desk tidy carefully. The first compartment is currently filled with sixty-seven steel paperclips. I can't remember when I bought these paperclips, or where I got them from, and the clips themselves offer no clues. I can only apologise for the vagueness of my records. But before you criticise me for this oversight, perhaps I am simply a product of my environment. As a supposed civilisation, we have been so blasé – so arrogant – that we haven't even bothered to keep a proper record of who invented the paperclip.

When you think of paperclips, you immediately think of a specific form – the familiar round-ended, double loop design. The wire trombone shape. But that's only one variety of clip; the 'Gem', which gets its name from a British company called Gem Limited, who, even if they weren't directly involved with the development of the clip, were clearly able to market it well enough that the name stuck. There are many different (and some might say better) types of paperclip. How can it be that we have no real idea of who invented it? One difficulty is that with so many different forms, so many different designs, there are just as many pretenders to the throne. Claims are advanced and myths develop. One common theory is that the paperclip was invented by a Norwegian patent clerk named Johann Vaaler in 1899. His patent application (filed in Germany in 1899 and then two years later in the United States) was for a clip made from 'a spring material, such as a piece of wire, that is bent to a rectangular, triangular or otherwise shaped hoop, the end parts of which wire piece form members or tongues lying side by side in contrary directions'. One of the illustrations he included in his application did in some way resemble a Gem, but as the Early Office Museum web site (my favourite place on the internet) so brutally puts it, 'his designs were neither first nor important'.

Vaaler's title as the supposed father of the paperclip was given to him posthumously. And as the story grew, it accidentally managed to turn him into a folk hero of sorts in Norway. During the years of Nazi occupation, the paperclip was worn as a symbol of resistance in Norway. This wasn't actually anything to do with Vaaler being Norwegian (even though Vaaler's original patent application had been rediscovered in the 1920s, the belief that he'd invented the paperclip didn't become widespread until later), but it was meant as a subtle sign – the binding action of the paperclip acting as a reminder that the Norwegian people were united together against the occupying forces ('we are bound together'). In the years following the war, belief that Vaaler had invented the clip began to spread. The story started appearing in Norwegian encyclopaedias and soon merged with stories of the resistance to elevate the paperclip into something approaching a national symbol. In 1989, the BI Business School erected a 7-metre-tall paperclip in Vaaler's honour on their Sandvika campus (this statue was later relocated to the Oslo campus). However, the statue is not actually of the same design Vaaler patented – it's a modified Gem (one end of the clip being slightly squared). Similarly, ten years later, when Vaaler was commemorated on a Norwegian postage stamp, it was a Gem which was shown next to his picture rather than the clip he actually designed (although a copy of his patent application was included in the background).

Closer to the Gem than Vaaler's design was a patent issued to Matthew Schooley in 1898 for his 'Paper Clip or Holder'. Schooley's design was an improvement on the other clips available at the time, as he explained in his patent application:

> I am aware that prior to my invention paper-clips have been made somewhat similar to mine in their general idea; but so far as I am informed none are free from objectionable projections which stand out from the papers which they hold.

Whereas Vaaler's design was a flat loop of wire, Schooley's design coiled round upon itself, allowing it to lie 'flat upon or against the papers which it binds together, presenting no projections or appreciable points upon which other objects may catch'. Furthermore, Schooley added, 'by its construction there is caused no puckering or bending of the papers'. It was an improvement, but it still wasn't quite a Gem.

The first time a recognisable Gem-type clip appears in patent literature is in 1899. William Middlebrook applied for a patent for a machine to automatically manufacture 'wire clips for binding or securing papers in lieu of pins' and included in the patent application is an image showing the 'general shape and character' of the clips the machine manufactured. The clip in this illustration is clearly a Gem-type design, but the clip itself wasn't part of the patent, it was just to show what the machine could do. However, the Gem was actually known for at least a decade before that. Professor Henry Petroski (author of *The Evolution of Useful Things*) cites an 1883 edition of Arthur Penn's *The Home Library*, which celebrates the Gem for its superiority over other devices for use in 'binding together papers on the same subject, a bundle of letters, or pages of a manuscript'.

While the anonymous inventor of the Gem pre-dated both Schooley and Vaaler, there are several even earlier paperclip designs. The most common recipient of the title 'inventor of the paperclip' is Samuel B. Fay, although it wasn't even paper that he had in mind when he developed his clip – his 1867 design was for a 'Ticket Fastener' intended to attach 'tags or tickets to fine fabrics to supply the place of pins, which have heretofore been used for that purpose, and which injure the fabrics to a greater or less extent by perforations' (although as he explained in his notes it could also be used to attach two pieces of paper together). Fay's clip consisted of a length of wire bent 'so as to form a loop at one end, or a bifurcated bar, the legs of which are then twisted or turned to cross each other, thus forming a spring clasp'. This design is more or less identical to the brass Premier-Grip Crossover Clips which fill the second compartment of my Velos desk tidy.

In his 1904 autobiography, the philosopher Herbert Spencer claimed to have invented a 'binding pin' as far back as 1846. Today best known for coining the phrase 'survival of the fittest', it seems Spencer was also something of an inventor. The device was intended to hold 'unstitched publications' such as newspapers and periodicals together for ease while reading (the newspaper 'being opened out in the middle, these binding pins, being thrust on to it, one at the top of the fold and another at the bottom, clipped all the leaves and kept them securely in their positions'). Spencer signed an agreement with Messrs Ackermann & Co. to produce and sell the binding pins. In the first year, sales of the pins made around £70 (the equivalent of £6,150 today) but sales quickly began to drop off after this. Initially, Spencer blamed Ackermann for failing to sell more of the clips ('I supposed the fault to be with Mr Ackermann who was a bad man of business, and who, failing not long afterwards, shot himself') although he later claimed it was the public's 'insane desire' for novelty which was so 'utterly undiscriminating that in consequence of it good things continually go out of use, while new and worse things come into use: the question of relative merit being scarcely entertained'. So much for survival of the fittest.

Prior to these devices, straight pins had been used for attaching papers; however, there are several obvious problems with the pin method. The main one being that pinning involves puncturing the paper. Whatever papers you wanted attached together are now attached together but they also now have holes in them. Hardly ideal. A system which avoids this is clearly an improvement. Also, anything without the sharp points of a pin would be kinder on the fingers. A clip, such as the one designed by Fay, seems, in retrospect, such an obvious improvement on the existing pins that you wonder why no one thought of it earlier. But this question – 'why didn't anyone think of it sooner?' – sort of misses a fundamental point in the process of design. Within its own ecosystem, the straight pin worked fairly well. Yes, there were problems associated with it, but without a viable alternative, there was no point in complaining. There was nothing to force the pin to develop. It was happy as it was. The ecosystem in which the pin lived needed to change before the pin could evolve. In the late nineteenth century, three things happened which changed that ecosystem and allowed a new species – the paperclip – to emerge.

Most obviously, for the paperclip to exist at all, you need the technology to reliably produce steel wire with the elastic properties which the clip requires in order to function successfully. Secondly, you need to be able to manufacture and sell these wire clips at a cost which is acceptable to the public (even though people may have been unhappy with their papers being pierced by straight pins, they were willing to accept this aesthetic assault as any alternative was too expensive to be practical). Finally, you had a burgeoning bureaucracy – a side-effect of the industrialisation which enabled the first two factors. It was the birth of the office environment and a new infrastructure was required. More paperwork necessitated some new method of organisation; the era of the paperclip was born.

As none of these forces was unique to any particular place, it's not surprising that during the later years of the nineteenth century a multitude of designs emerged in various countries more or less simultaneously. From 1867 onwards, a bewildering number of patents were applied for by a host of inventors all hoping to find the best way of using a single piece of metal to attach two or more sheets of paper together. These alternative clips took many forms. There was the 'Eureka' clip, a sort of segmented oval shape, cut out of a sheet of metal with a central prong to hold the papers together, patented by George Farmer in 1894; the 'Utility' clip from 1895, which was similar to an old-fashioned ring-pull, folded back on itself; the 'Niagara', which was basically two of Fay's clips joined together, patented in 1897; the 'Clipper', a pointy version of the Niagara from the same year; the 'Weis' clip, an equilateral triangle within an isosceles, patented in 1904; the spectacularly named 'Herculean Reversible Paper Clip' where the wire was bent into two slightly wonky isosceles triangles; the 'Regal' or 'Owl' clip, which sort of looked like an owl, if the owl had been raised in a rectangular cage which was too small for it and grew up deformed into a weird boxy shape; and the 'Ideal' clip, a complex butterfly-shaped wire arrangement, patented in 1902. The list goes on and on: the 'Rinklip', the 'Mogul', the 'Dennison', the 'Ezeon'.

A man named George McGill submitted around a dozen patent applications for new paperclip designs between 1902 and 1903. He must have been a restless man, obsessed with stationery (he also designed paper fasteners, ticket holders, and staplers). I imagine him constantly doodling new designs on the back of envelopes and scraps of paper, a frustrated wife despairing as she suspects that while they're lying in bed together or having dinner, part of his mind is constantly elsewhere, searching for the perfect wire clip design. How close did he get to realising his dream? It seems he had limited success. The Early Office Museum specifically limit their collection of early paperclip designs to those registered before 1902 because of the chaos caused by McGill:

> We did not include paper clips that were patented after 1902 unless we could find evidence that they were produced. We used that cut-off date because 13 paper clip patents were awarded in 1903, 10 of them to one inventor, George W. McGill. With the exception of McGill's design for the Banjo paper clip, we have found no evidence that any of these was produced or advertised.

It may well be true that many of McGill's designs did not progress beyond the registering of a patent, but the Early Office Museum are being a bit unfair on the man. At least one other of his 1903 designs went into production – I have a box of Ring Clips ('Patented to Geo. W. McGill, June 23 & Nov 17 1903') on my desk.

Despite this period of wild experimentation, the narrow double loop form of the Gem has remained the most enduring paperclip design. Often cited as an example of 'perfect' design, the clip has featured in exhibitions at the Museum of Modern Art in New York and the Vitra Design Museum in Germany. Emilia Terragni, one of the editors of the *Phaidon Design Classics* series, named the paperclip as one of her favourite objects:

> Because in the paperclip you have the essence of design: you have beautiful design; you have a simple mechanism; you have something that's never changed in a hundred years – it's still the same. It's still very functional and everybody uses it.

But, is the Gem paperclip really as perfect as so many claim? In articles about the beauty of the clip's design, it is always shown in isolation and never shown in use. Once it is actually used to attach papers together, half of the classic form is hidden. If the clip is used to hold together a particularly thick document, it can become distorted and bent out of shape. In many ways, its functional qualities have become overstated as the simplicity of its design has become cherished.

The claim that the design has not changed in a hundred years is also questionable. It's true that the paperclips available today are very similar to those illustrated in advertisements from the 1890s. But there are also lots of paperclips that share many of the characteristics of the Gem, but which have been given subtle tweaks here and there. You can get 'lipped' clips, where the bottom of the inner loop is raised, to allow the clip to slide on more easily (although this idea has been around since George McGill's patent of 1903). Another variation is the 'Gothic' clip, designed by Henry Lankenau in 1934 – whereas the classic Gem has rounded Romanesque ends, the Gothic clip has a squared top allowing it to lie flush with the top of the paper and a pointed bottom, making it easier to attach. Corrugated clips provide extra friction, preventing the clip from sliding off so easily. The differences are slight, but changes have been made. From a recent trip to Ryman, I'd say that the paperclips were split about 50/50 between pure Gem-like clips, and modified versions.

So why is there this prevailing belief that the Gem represents perfection, when the reality is that it's actually not quite as good as everyone thinks? The belief seems to come from the fact that on almost every level the Gem is more than satisfactory. It's not perfect, but it's good enough. 8/10 all round. When a tweak is made, the new design performs better in some ways, but then worse in others. The 'lip' makes it easier to slide on, but can make a pile of documents bulkier. The 'Gothic' clip is also easier to use, but the pointed end can scratch or tear papers. The corrugated clip is less likely to slide off accidentally, but is then more fiddly to remove. The Gem isn't perfect, and people will continue to try to improve it, but the struggle will be to find a new design that is as balanced as the Gem.

As well as the desk tidy itself, the Velos box also included a small leaflet listing the other products in the same range, including a series of office basics:

130 Stamp Rack
176 Carousel Desk Tidy
006 Twin Roller Damper
1365 Damper
1502 Moistener Stamp Pads

There was a selection from the Velos range of Staplers & Staples:

347 Long Arm Stitcher
300 Falcon
325 Windsor
330 Tacker

23	Staple Remover
321	Snipe

They also had perforators and hole punches:

4362	Heavy Duty Punch
4363	Easy Punch
950	Eyeletter & Punch
4314	Lightning
4316	Heavy Duty
4324	Four Hole

The Velos range boasted over seventy-five different elastic band sizes and five sizes of thimblettes. Desk tray units were available with three or five tiers and in 'swivel' or 'riser' arrangements. There were six desk pencil sharpeners and three 'pocket models'. Map pins were available in over twenty colours and could be provided in tubes or blister packs. There were three sizes of cabinets for storing micro-fiche sheets and index cards (5' × 3', 6' × 4' and 8' × 5'). Each set of products was shown photographed against a brightly coloured background. They all had that peculiar gloss you see in colour supplements from the time. Everything appearing thickly shiny, looking like it's been dipped in treacle. The items themselves displaying that period's insatiable obsession with orange and brown as a colour scheme. A combination which somehow makes me nostalgic for a time I am too young to really remember.

The third compartment of my Velos 1377 Revolving Desk Tidy is filled with brass-plated drawing pins. As its name suggests, the 'drawing pin' was originally used by draughtsmen to hold down the drawings they were working on. These pins would have had different shapes and designs, having evolved from simple straight pins. As with the development of the paperclip, there is some debate over who exactly invented the drawing pin as we know it today. Some credit the Austrian engineer Heinrich Sachs with the invention. The Sachs pin, designed in 1888, consists of a small steel disc, into which a narrow V-shaped nick is punched. This nick is then bent back to form the point of the pin. While this design is not common in the UK, it remains popular in other parts of the world.

The more common form of drawing pin, and the variety which I keep in my desk tidy, is the brass version often called a 'thumbtack' in the US. A small brass domed head to which a short spiked point is attached. Some claim the pin was invented by a German watchmaker named Johann Kirsten sometime between 1902 and 1903. One theory is that prior to this, Kirsten (like many before him, no doubt) used a simple straight pin to hold down his drawings as he worked. Realising that a pin featuring a large, flattish head would be kinder on the thumb, he beat out a small brass disc and punched a nail through it. However, it wasn't Kirsten who benefited from his design. While Kirsten was able to sell a small amount of the pins to other local craftsmen, he still found himself short of cash (possibly the result of his heavy drinking – he was supposed to have once ordered a carriage to take him from his house to the pub next door, while his children sat at home starving) and was forced to sell the design to factory owner Arthur Lindstedt. Unfortunately for Lindstedt the pin had a design flaw; the head would come away from the nail when pressure was applied. This severely limited its commercial potential; it just wasn't up to the job. When Arthur's brother, Otto, took over the company, he asked his staff to solve the problem and it was this reworked pin design which Otto registered at the patent office in Berlin (patent number 154 957 70 E) on 8 January 1904. The pin made Lindstedt a fortune, with each worker at the Lindstedt factory producing thousands of pins each day for export all over Europe (Adam Smith would be proud). Kirsten was soon forgotten about.

Well, perhaps not entirely forgotten. In 2003, Christa Kothe, owner of a small hotel just outside Lychen, paid to have a statue built to mark one hundred years of the drawing pin.

The statue was situated directly outside the hotel, rather than in central Lychen or on the site of Kirsten's workshop. Some might say this suggests the whole thing was just a publicity stunt for the hotel rather than a truly altruistic celebration of this apparent hero of stationery. But the memorial wasn't just on the wrong side of town, it was in the wrong country – and if it was intended to mark the 100th anniversary of the pin's invention, it was also several decades too late, as Kirsten's pin was far from the first.

The Oxford English Dictionary defines the 'drawing pin' as 'a flat-headed pin used to fasten drawing-paper to a board, desk, etc' and quotes an 1859 text (F. A. Griffiths's *Artillerist's Man*) as an example of usage:

> Fixing it firmly by means of drawing-pins ...

But we can go further back than that. The third volume of *The Register of the Arts and Sciences* (1826) includes this mention of drawing pins:

> If a small fine drawing pin be therefore fixed in that place, and into the centre of the circle struck, the radii may be drawn with great facility and exactness.

This passage doesn't really make clear what shape the 'drawing pin' referred to was. It could have simply been a variation on the straight pins which had been around for years; the description of the pin as small and fine suggests something like this. In the passage the pins are being used in a slightly different way – to help draw a smooth curve rather than pinning down the drawing, so it's possible that the pin was nothing like the sort of drawing pins we are familiar with today. Perhaps, therefore, the Johann Kirsten story shouldn't be immediately dismissed. However, there's a much more specific reference in Robert Griffith Hatfield's book *The American House-Carpenter* from 1844:

> A drawing-pin is a small brass button, having a steel pin projecting from the underside. By having one of these at each corner, the paper can be fixed to the board.

A similar description appears in John Fry Heather's 1851 *Treatise on Mathematical Instruments*:

> The drawing pin consists of a brass head with a steel point at right angles to its plane.

And just to make clear that the pin referred to here is the same as the 'drawing pin' we know today, Heather even includes an illustration. It is indeed the drawing pin we know and love. Poor old Johann. At least he can take comfort in the fact that, just like the other Johann, thought to have invented the paperclip, his fellow countrymen were moved enough to build an inaccurate statue in his misremembered memory decades after he died. What an honour.

Though the drawing pin seems to be of European heritage, the forth compartment of my Velos 1377 Revolving Desk Tidy is filled with its US counterpart – the 'push-pin', designed by Edwin Moore in New Jersey in 1900. Working in a photographic laboratory, Moore was searching for an easy way to hang up prints as they dried, as he was dissatisfied with the pins currently available:

> I have, however, found that the use of such [an] article is attended with many disadvantages. The body portion presents no means for firmly holding the same between the fingers when being inserted, and consequently the fingers slip and tear or blotch the film. Furthermore, the water employed upon the film corrodes the pin and the metallic cap, thus causing stains upon the film.

Moore's solution was, as he described it, simply 'a pin with a handle'. A short steel spike embedded in a miniature top hat made of glass. In one variation, he suggested tapering the end of the head so it could then be 'suitably ornamented', and illustrated it with a tiny model of a sort of pig/dog/bear head (the drawing isn't very clear). Leaving the photographic industry, Moore went into business manufacturing these 'push-pins' with $112.60 of private capital. He would work making the pins in the evening, and then sell what he'd made the next morning. His first order was for one gross, which he sold for $2. Fortunately, bigger orders started flowing in, and before long he received a $1,000 order from the Eastman Kodak Company. Reinvesting the money he received from these orders, Moore started heavily investing in advertising to promote his products (his first advertisement appeared in a 1903 edition of *The Ladies' Home Journal*, and cost him $168) and the company rapidly grew. In fact, the Moore Push-Pin Company continues to trade to this day, manufacturing 'little things' such as numbered and round-headed map pins (long thin pins with a ball-shaped head – these pins are in the fifth compartment of my Velos desk tidy); Pic-Sure-Stay® and Snub-It™ photo frame hangers; Tacky-Tape®; and, of course, push-pins themselves (although sadly, they no longer sell the original glass variety). The range now consists of plastic, aluminium and wood, as well as the new Thin Pin™ (a sort of flattened version of a push-pin which can be bent by 90° to clip items without piercing them).

The push-pin has several advantages over its European rival. The 'handle' means the pin is a lot easier to remove than a drawing pin, particularly if the drawing pin has been pushed fully flat against the surface to which it's been pinned. A 1916 edition of *Popular Mechanics* describes an attempt to combat this problem, with a new type of drawing pin featured:

> ... a semi-circular handle, the ends of which are turned in to engage holes in the sides of the head. One half of the head may be made of a smaller radius than the other half, thus permitting the handle to be turned down flat and thereby completing the circle of the head.

This seems unnecessarily complicated and so it's no great

surprise that the 'drawing pin with a handle' never took off, particularly when the push-pin solved the same problem with much less fuss (although the shape of the push-pin means the head doesn't lie flat against the surface it has been pinned to and so makes it unsuitable for, say, a notice board in a narrow corridor, where a careless shoulder could bring a sheet of A4 fluttering to the floor with potentially trivial consequences).

Another advantage the push-pin has over the conventional drawing pin is that it reduces the risk of one of the most common stationery-related injuries. A dropped drawing pin can often land with its spike pointing upwards, just waiting to embed itself in the sole of your foot (followed by the confused howl of pain as a foot plunges down on the pin and the pathetic hop to a nearby chair, as the foot is lifted to see the pin hanging limply out of the flesh). A dropped push-pin, however, with its smaller head and longer body, is much less likely to land in the same way; it will land on its side instead. If only the proprietors of RC Hammett Butchers Limited had used push-pins, perhaps Mrs Doris Nichols of South Chingford could have avoided her ordeal. On 18 June 1932, Mrs Nichols bought a chicken and five pork pies from her local butchers. Later that evening, as she merrily tucked into one of the pies, she felt a sharp pain in the back of her mouth. She reached into her mouth and pulled out a drawing pin. Her throat became inflamed and, even after seeing her doctor, she was still unable to eat or drink. 'On June 22 she became very sick,' explained a report in *The Times*, 'and on June 23 and 24 vomited blood.' There then follows what must be one of the most horrifying sentences I have ever read about drawing pins: 'On June 25, she passed out a drawing pin.'

The butchers admitted liability, but claimed that the accident 'might have happened anywhere', explaining that a man had 'removed a piece of American cloth from the table and left the drawing-pins which had fastened it down lying there, and the person bringing in the pork pies had inadvertently placed them on the pins'. Understandably, the ordeal took some getting over, and during the next few months, Mrs Nichols lost almost two stone in weight. At a court case in December that year, her doctor, Dr Bryan Buckley Sharp, explained that she had 'had an alarming experience which would be liable to upset a highly strung woman'. In giving judgment, Mr Justice MacNaghten awarded Mrs Nichols £200 in damages, saying he 'could not imagine anything more uncomfortable than to swallow drawing-pins, and anyone doing so might find it extremely difficult to forget about'; however he accepted that it was 'evidently a pure accident' and that 'there was nothing wrong with the pork pies'.

The final compartment of my Velos 1377 Revolving Desk Tidy is filled with two dozen stainless steel clips for a Rapesco Supaclip dispenser. The Rapesco Supaclip dispenser is a sort of stationery-related cross between a Pez dispenser and a Sonic Screwdriver; a small, transparent handheld device operated by a spring-loaded thumb trigger. The metal clamp-like clips are forced open through the mouth of the dispenser and then snap down, firmly clipping the papers together. The clips can then be removed by hand and reused. Rapesco describe the Supaclip as 'the original and best', and appear convinced that there is some sort of widespread attempt to copy the magnificence of the Supaclip dispenser:

> Say goodbye to the paperclip and beware of imitations. Supaclip® 40 Dispenser offers maximum 40 sheet capacity – other copies don't.

Perhaps their paranoia is justified. One question which Rapesco are apparently asked so frequently that they've been forced to include it on their FAQ page is:

> Q Will your Supaclips fit other brands of dispensers?
> A They may fit, but will not operate consistently.

The Velos trademark was registered by Rees Pitchford & Co. Ltd on 14 March 1946 and covered the following goods:

> Adhesive materials (stationery) not primarily for use in photographic mounting; and artists' brushes, office requisites and appliances (other than furniture) and printers type, but not including knives, pliers, or punches or any goods of the same description as any of these excluded goods.

But the brand had actually been established for a while before that. Rees Pitchford & Co. were originally Frank Pitchford & Co. and were founded in the early years of the twentieth century. By the late 1930s, the company had changed its name to Rees Pitchford & Co. Ltd, and the Velos brand thrived for many years, with the Velos 'V' logo proudly stamped across staplers, pencil sharpeners and hole punches. However, like so many other stationery brands, eventually it succumbed to corporate might and in 2004 the trademark was assigned to ACCO Brands.

The anonymous-sounding ACCO Brands Corporation is one of the largest suppliers of office products in the world, slowly swallowing other companies and amalgamating their brands into the ACCO portfolio. Starting out as the American Clip Company in 1903, the company includes Wilson Jones ('inventor of the 3-ring binder', formed in 1893), Swingline ('the #1 brand in stapling and the leader in workspace productivity with products such as staplers, punches, and trimmers', formed in 1925), General Binding Company ('the global leader in binding and laminating equipment and supplies', formed in 1947), Rexel ('with 70 years of great design and innovation, the Rexel product portfolio ranges from shredders, trimmers and an extensive range of filing products to desk accessories and desktop tools') and Derwent Pencils ('We've been making pencils in Cumbria since 1832, and we think we've perfected the art') among others.

So Velos are no more. Absorbed into a faceless multinational. The name lives on, just about. Rexel use the name for their range of eyelet punches, but in doing so, they have moved from being a company producing countless office basics and stationery essentials to one which provides tools for haberdashery. I'm not interested in haberdashery. I'm interested in stationery. But does it even matter? I'd never heard of the brand until I found an old box in a shop in Worcester Park. Why should I care about their history? But the more I thought about Velos, the more I thought about other companies. I thought about companies I'd never even heard of. If there was Velos, who else was out there? This, in its own small way, is part of our cultural heritage and names that were once well known have disappeared, barely leaving any sign that they existed in the first place. Which names, familiar to us today, will fade into obscurity tomorrow? But more than that, I thought about people. The people behind these objects that we take for granted. The names behind the brand names. Their lives, their histories. Who were they? What were their stories? I wanted to find out.

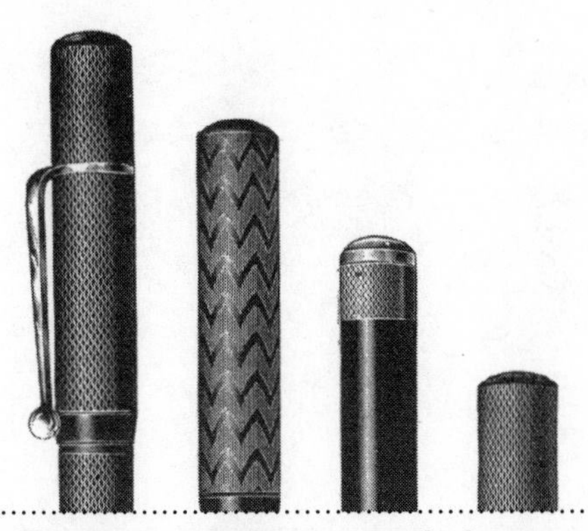

CHAPTER 2

Everything I know about people I learnt from pens

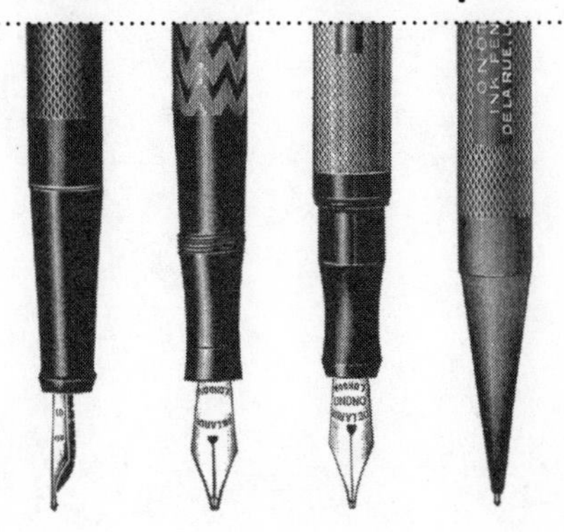

Before I started writing this book, I signed a contract with my publisher. That's how it works. It's not that we don't trust each other, it's just better for everyone concerned if there's a contract rather than just a handshake and a nod. Considering the fact this is a book about stationery, I felt under pressure to make sure I used the right pen for the signing.

But what is the right pen? I gave it a lot of thought. At first I considered a fountain pen. Something mature. Grown-up. Signing my name in royal blue ink. But that seemed too ostentatious, and besides, I've always found them a bit scratchy to write with. Using a fountain pen wouldn't be true to who I am. A gel pen, with their brightly coloured inks, seemed to be too trivial. The more I thought about it, the more convinced I was. There was only one option. Others may have gone for something flashy or expensive, but not me. I wanted something humble, yet iconic. Something definitive. Something extraordinary in its simplicity. The BIC Cristal.

For many people, the BIC Cristal is synonymous with the ballpoint pen; a pen so familiar that most of us don't even know its proper name. But to call it a 'BIC Cristal' feels like an affectation. To millions, it is simply a 'Bic Biro'. However, the histories of BIC and Biro are the histories of two very different companies, suing and counter-suing each other until buyouts and marriage finally brought them together.

Marcel Bich launched the BIC Cristal in France in 1951. Bich had moved to France from Italy in the early 1930s. He worked alongside Edouard Buffard at an office supplies company owned by the English company Stephens Ink, and following the end of the Second World War, he and Buffard bought a small workshop in Clichy, just outside Paris. With Buffard as production manager, Bich became managing director of *la société Porte-plume, Porte-mines et Accessoires* (PPA) and the company began producing parts for local fountain pen companies. Towards the end of the 1940s, *la société PPA* began to receive enquiries about a new type of pen which had recently been developed: the ballpoint. Bich quickly became aware of the potential of this new type of writing instrument and decided to develop a design of his own.

During his time at Stephens Ink, Bich had developed a friendship with businessman Jean LaForest. LaForest owned a small pen company and, in 1932, he co-registered a patent with his colleague Jean Pignon for a ballpoint mechanism. Bich had identified that the problem with a lot of ballpoint pens on the market at that time was to do with the ink, which would leak and smear on the page, or dry up in the barrel. Working with another company, Gaut-Blancan, Bich developed an ink which would be suitable for this new type of pen. Building on the work of LaForest and Pignon, Bich was able to develop a pen which overcame many of the problems faced by his rivals who were also moving into the ballpoint market. Modelling the barrel of the pen on the hexagonal body of a traditional wooden pencil, PPA's *Décolletage Plastique* design team created the now familiar design. By the end of 1950, the new ballpoint was ready to be released.

The 'h' from the end of Bich's surname was dropped to

form the now familiar 'BIC' brand, and the pen was originally sold by *la société PPA* (it would be another two years before the formation of *la société BIC*). Available in five colours, including not only the commonplace black, blue, red and green, but a violet as well – for, I don't know, special occasions, there were three different pens in the BIC range; the disposable *Cristal* (60 old francs – around £1.50 today), the refillable *Opaque* (100 old francs or £2.50 today) and the luxury *Guilloché* (200 old francs, about £5 today). Despite its relatively high cost compared to the refillable options, it was the convenient and disposable Cristal which really appealed to people; in the first year alone, over 25 million units were sold.

In the next few years, the company began an aggressive marketing campaign to promote the pen; advertising on radio, in the press and in cinemas. During the 1952 Tour de France, the company hired a van with a giant BIC Cristal pen attached to the roof to follow the cyclists along the route. With huge crowds lining the streets to watch the race, the van was what would now be described in advertising terms as 'prime real estate' and the campaign clearly worked; by 1958, the company was producing a million pens a day, and BIC has continued its association with the Tour ever since.

Originally, the writing ball in the pen was made from stainless steel, but this was switched to tungsten carbide in 1961. The switch allowed the company to produce a pen with a finer point. In order to differentiate between the standard 1 mm point pen and its latest 0.8 mm Cristal Fine pen, the company gave the new model a casing to match its recently adopted corporate colour: bright orange. This colour is still used today and has been adopted by other pen manufacturers to distinguish fine line pens from their thicker cousins. To advertise the new tip, the company hired graphic artist Raymond Savignac to create a mascot. The 'BIC Boy' character, based on a schoolboy holding a pen behind his back (a schoolboy who happens to have a tungsten carbide ball for a head), is still used to this day.

The design of the BIC Cristal was based on that of the traditional wooden pencil, but its ancestry can be traced much further back; all the way to the beginning of human civilisation itself. For as long as 30,000 years, humans have made markings on walls and in clay as a way of understanding the world around them. The earliest cave drawings would have been made simply with a finger in clay, but as these drawings became codified and formalised into something approaching language, simple tools would be used to produce pictographic symbols. Pieces of reed would be pressed into clay to form cuneiform script (this early system of writing was developed by the Mesopotamians in around the fourth century BC, and takes its name from the Latin word *cuneus* meaning 'wedge'). In Egypt, reed brushes would be used with ink made from soot and water on sheets of papyrus. Gradually, this reed brush was replaced with a reed pen – created by shaping a piece of reed to a narrow point and then making a small cut to create a split nib. Ink could be poured in through the top of the hollow reed, which would seep down through the pen to the nib, similar to a modern fountain pen.

From around the sixth century, feather quills began to emerge. The broad line produced by the reed pen was acceptable for writing on a rough surface like papyrus, but the development of smoother materials such as parchment and vellum needed a finer line. The flexible quill of a feather (often goose) meant it could be cut to a much sharper point, and was less likely to split than fibrous reed. The earliest specific reference to the quill comes from Saint Isidore of Seville in 624, who describes both the *pinna* (quill pen) and *calamus* (reed pen), showing that both continued to be used side by side:

> The scribe's tools are the calamus and the pinna. With these, words are fixed onto the pages. The calamus is from a plant, the pinna from a bird. Its tip is split into two, while the whole shaft preserves its unity.

Isidore also explained how the two writing instruments got their names:

> The calamus is so called because it places liquid, whence among sailors 'calare' means 'to place'. The pinna is so called from 'hanging' [pendendo] that is 'flying', for it comes, as we have said, from birds.

The feather quill's suitability for use as a writing instrument is illustrated by the fact that it remained in use until the nineteenth century, when it was eventually replaced by the metal nib. Pens with metal nibs had been around since Roman times but were relatively rare, as it was difficult to manufacture a thin metal nib which could produce lines of the same delicacy and expressive quality as the quill. One advantage the reed and these primitive metal pens had over the quill was that they could hold a small reservoir of ink inside them. Constantly dipping a quill into a pot of ink not only made the writing process slower, but could also produce inconsistency in the strength of the line.

In the tenth century, caliph Al-Mu'izz li-Din Allah commissioned the construction of a metal pen, considered by some to be a prototype of the modern fountain pen. Recorded by historian Qadi al-Nu'man al Tamimi in his 962 work *Kitab al-Majalis wa'l-musayarat wa'l-mawaqif wa'l-tawqi'at* ('Book of homiletic sessions, accompaniments on journeys, halting places and administrative decrees'), the caliph described his desire to create 'a pen which can be used for writing without having recourse to an ink holder and whose ink will be contained inside it'. The pen would be filled with ink, and when the user has finished writing with it 'and the ink has become dry, the writer can then put it in his sleeve or anywhere he wishes and it will not stain it at all, nor will any drop of ink leak out of it. The ink will only flow when he expressly desires it to do so and when there is an intention to write it'. When asked by Qadi al-Nu'man if such a thing was possible, Al-Mu'izz replied, 'It is possible, if God so wills'.

Within days, the caliph's craftsman had prepared a pen 'fashioned from gold', but because it 'released a little more ink than was necessary', he ordered for the pen to be adjusted. The new pen could be 'turned upside down in the hand and tipped from side to side, and no trace of ink appears from it'. The Qadi was clearly moved by the pen, writing that he saw in it

> a fine moral example, in that the pen does not release its contents except when specifically requested to do so and for some useful purpose which is part of the original reason for asking it to write. It only bestows benefit on a person really desiring it, and it does not let its ink flow except for a person who has a right to summon it because the pen approves of him.

Sadly, Qadi al-Nu'man did not include any description of how the pen was able to make these judgements of character, and no details of the pen's construction have been found.

Throughout the sixteenth century, attempts were made to develop a pen with a self-contained reservoir of ink. In Leonardo da Vinci's *Codex Atlanticus*, there are drawings from 1508 which show pens with cylindrical containers of ink inside, sealed with a cap to prevent the ink from leaking. Gustav Adolph II, king of Sweden, received a silver pen in 1632 with an ink reservoir which allowed the pen to be used for up to two hours before being refilled. Published in 1636, Daniel Shwenter describes in his *Deliciae Physico Mathematicae* ('Delights of Mathematical Physics') a quill pen with a second quill inserted inside. This second quill would be filled with ink and sealed with a piece of cork. In 1663, Samuel Pepys refers to receiving a letter from William Coventry, 'and with it a Silver pen he promised me, to carry inke in; which is very necessary'. Indeed, it may well have been this pen which Pepys used at London Bridge when he 'did 'light at the foot of the bridge, and by helpe of a candle at a stall', write a letter to Tom Hater commenting that he 'never knew so great an instance of the usefulness of carrying pen and ink and wax about one'.

By the early eighteenth century, metal-nibbed 'endless' or 'continuous' pens, capable of around twelve hours of use, began to be developed. But the relative complexity of the designs, as well as the risk of the ink leaking and high cost meant that the quill remained in common usage until the middle of the nineteenth century, when it was eventually replaced with a metal version of itself – the dip pen.

During the nineteenth century, improvements in manufacturing meant that nibs could be produced with much finer and more flexible points than had ever been possible before, and so metal nibs began to grow in popularity. The metal nibs were much longer-lasting than quill tips and could be mass-produced cheaply. The nibs would be attached to rosewood or silver pen-holders, with the nibs easily replaced once they wore out. But some disliked the scratchy nature of the sharp metal nibs. Victor Hugo dismissed them as 'needles', and the French writer and critic Jules Janin called them 'the true root of all evil', saying:

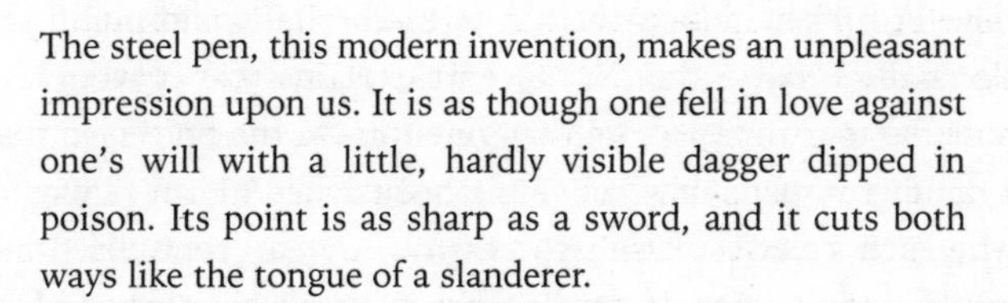

> The steel pen, this modern invention, makes an unpleasant impression upon us. It is as though one fell in love against one's will with a little, hardly visible dagger dipped in poison. Its point is as sharp as a sword, and it cuts both ways like the tongue of a slanderer.

Sadly for Janin and Hugo, the quill was on the way out.

In 1809 Peregrine Williamson patented his design for a 'metallic writing pen' in Baltimore, but it was Birmingham, England that would become the worldwide capital of steel pen manufacture. John Mitchell developed a system to mass-produce steel nibs in 1822, and six years later, Josiah Mason opened his factory – soon Mason would become the biggest single manufacturer of pens in the country. By the middle of the century, over half of all the steel pen nibs produced worldwide were made in Birmingham. The low cost of these mass-produced nibs, as well as their reliability, meant that they were popular in schools, where they would continue to be used well into the second half of the twentieth century (even in my primary school in the 1980s, the desks in some of the classrooms still had inkwells). Steel-nibbed dip pens may have replaced the quill, but they still had the same drawback: the pen needed to be dipped into a pot of ink after a few strokes.

Early fountain pens were commonly refilled using an eyedropper or pipette, transferring the ink from a bottle into the pen barrel. The eyedroppers consisted of a thin glass tube, attached to a rubber bulb. Not only did this mean that you always had to have the eyedropper to hand, but the glass of the eyedroppers was very fragile and could easily break. The eyedropper was slowly replaced by 'self-filling' pens which included a rubber sac ink container.

One mechanism for refilling fountain pens which sadly never caught on was that suggested by Germany's Hugo Siegert in 1892. With his 'Combined Inkstand and Penholder', the pen would be attached to a large bottle of ink by a long rubber tube. A rubber bulb at the top of the bottle would pump ink along the tube to the pen. Siegert explains that it would be possible to 'connect more than one tube to the bottle, so that a number of pens may be simultaneously fed'. I can't imagine why such a device wasn't successful. A giant, centralised ink bottle simultaneously feeding the pens of an office full of people via a network of long rubber tubes would look amazing; like something from Terry Gilliam's *Brazil*. If you work in an office, please suggest this to your boss or whoever orders the stationery. We need to make this a reality.

It wasn't until 1884 that the first commercially successful fountain pen was introduced: the Ideal, designed by Lewis Edson Waterman. Born in New York in 1837, Waterman worked as a teacher, book salesman and insurance salesman despite having received a fairly basic education. It was while he was working in this last job that he is believed to have been inspired to develop his improved fountain pen. As he was about to sign a major insurance contract with a client, his pen leaked, creating a large inkblot on the crucial document. In the time it took Waterman to get a new contract prepared, the client had gone elsewhere. Waterman became determined never to let that happen again.

As seductive as this story may sound, it is almost certainly completely false. David Nishimura, of the Vintage Pens web site, has studied the Waterman Pen Company's promotional literature and can find no trace of the 'inkblot' theory until 1921 (twenty years after Lewis Waterman's death), despite a lengthy article in the company's own *Pen Prophet* magazine in 1904, which went into considerable detail about the origin of the company. Nishimura believes the story was created by Waterman's advertising department to promote the image of Lewis Waterman as a kind of aw-shucks Jimmy Stewart down-to-earth everyman.

Waterman's pen was designed to 'secure and automatically regulate a certain and uniform flow of ink' to the nib. The pen featured a rubber barrel to contain the ink and a series of 'narrow slits or fissures' in the nib to draw the ink down through a combination of gravity and capillary action. Compared to many of the earlier designs, Waterman's idea was extremely simple. 'The feed of the Ideal is the essence of simplicity,' explained James Maginnis in his 1905 lecture on fountain pens, printed in the *Journal of the Society of Arts*. The simplicity of the design and the quality of manufacture meant the pen quickly took off. Within a couple of years, production had increased from thirty-six pens a week to a thousand pens a day. In 2006, Lewis Edson Waterman was inducted into the US National Inventors Hall of Fame. The brief biography on his page on the Hall of Fame web site says that Waterman 'is said to have vowed to invent a better writing instrument when an inferior pen leaked on an important insurance contract, delaying him long enough that he lost the client'. Aw shucks.

In 1913, the Parker Pen Company introduced the 'button

filler' system. This simple mechanism featured a button at the top of the barrel, which would be depressed as the nib was dipped into ink. A bar inside the pen would compress the internal sac and once released, the pen would be filled. Parker also introduced a new type of lid – the 'Jack Knife', which featured a sort of lid-within-a-lid to prevent leakage. 'It has the novelty that attracts,' wrote Parker, 'and the usefulness that closes sales.'

The majority of fountain pens produced by the Parker Pen Company at this time were fairly sober-looking (mainly made from black ebonite) but an observation by one Parker employee soon changed this. In 1920, Lewis Tebbel approached George Parker with a simple idea: rather than creating a product designed for the standard office worker, why not aim a bit higher? One story has it that Tebbel and Parker were looking out from the top of their office building when Tebbel began pointing out the limousines in the traffic below. Yes, the economy was still in trouble, but if some could afford to travel in luxury, there was also a market for luxury pens. Pens you could charge extra for.

The 'Big Red' Parker Duofold was the pen Tebbel had in mind. Produced in a bright red-orange ebonite, and deliberately positioned with a higher retail price than the rest of the Parker range, the pen quickly became a highly desirable status symbol. Further colours were added (including 'Modern Green', 'Mandarin Yellow', 'Jade Green' and 'Sea Green Pearl') whose names seem to emphasise the pen's luxury status. However, the pen was superseded by the Vacumatic, launched in 1933. This pen had almost twice the ink capacity of the Duofold, and featured a barrel with alternating bands of light and dark celluloid through which the level of ink could be viewed. But it would be Parker's next pen which would become their most successful.

Launched in 1941, the name of the Parker 51 was both forward-looking ('Ten years ahead') and nostalgic (the pen had been developed in 1939, the fifty-first anniversary of the Parker

Pen Company). Ten years earlier, Parker had launched their own quick-drying ink, Quink, but the company later developed an ink which dried even more quickly ('Writes dry with wet ink') and was available in a range of strong colours ('India Black', 'Tunis Blue', 'China Red' and 'Pan American Green'). Unfortunately, this new ink was corrosive and would attack the rubber ink reservoirs and celluloid barrels of most normal fountain pens on the market at that time. And so Parker developed the 51 with a body made of Lucite (a strong transparent plastic, then used in aeronautics) which could resist their otherwise unmarketable ink. The tip of the pen was made from Plathenium (an alloy of platinum and ruthenium attached to a 14K gold nib). Its 'all-precious metal point' meant that the point would wear in 'to your style of writing, polishing itself after a few hours of use to supreme smoothness – and then stays that way for decades and decades'.

> Now – if a two-hundred-pound man should take a new '51' Pen and boldly write with it for a few hours, it would break itself in to his hand in a decidedly masculine way. However, if a slip of a girl were to pick it up, she would be almost sure to break it in in a characteristically feminine way. It wouldn't take much of a magnifying glass to reveal which was which.

Designed by Marlin Baker, Gaylen Sayler and Milton Pickus under the direction of George's son, Kenneth Parker, the streamlined body of the 51, resembling a rocket or bomber aircraft (the similarly named Mustang P-51 aircraft had no connection to the pen, although Parker would emphasise the resemblance between the two in their advertising), and its hooded nib meant the design became an instant classic, catching the eye of former Bauhaus teacher László Moholy-Nagy who described it as 'one of the most successful designs of small utility objects in our period' and praised it for being 'light, handy, extremely well-shaped, unobtrusive and perfectly functional'.

As the US entered the Second World War, production of the Parker 51 was restricted by the War Production Board ruling to allocate scarce materials to the war effort. However, unlike their competitors, Parker continued to advertise heavily, creating demand for the product which would take several years to fulfil. 'Parker 51 pens now have to be rationed among dealers,' Parker advertisements explained. 'Yet you probably can place a reservation with one of the stores you usually patronize.'

While László Moholy-Nagy admired the Parker 51, another László wasn't so impressed with fountain pens. The son of a Jewish dentist, László Biró was born in Budapest in 1899. Following the outbreak of the First World War, Biró enrolled at an officers' school in 1917. After he was discharged from the army at the end of the war, Biró followed his older brother György by studying medicine. While at university, Biró became interested in hypnotism and co-authored a number of papers on the subject with his brother. Biró gave practical demonstrations and lectures, leaving university before completing his degree ('I was the first person in Hungary to deal seriously with practical hypnosis,' he would later write. 'I made so much money out of it that I lost all interest in continuing my medical studies.')

In the years that followed, Biró would try his hand at a variety of different jobs, never staying in one position for very long. He worked as an insurance salesman, as a book publisher, and for a company importing oil. While working for the import company, Biró made enough money to buy a second-hand Bugatti sports car from a friend, which he planned to enter into a race in Budapest two weeks later (despite the fact that he couldn't actually drive). As he began to learn how to drive, he struggled with using the clutch to change gear and decided to develop an automatic gearbox. This wasn't Biró's first invention; he had earlier patented a 'water fountain-pen' based on a design of his father's, which fed water through a tube to dissolve a solid ink core, as well as an early type of washing machine. Neither of these inventions appears to have had any success, however – he was too busy jumping from one thing to the next to pursue any particular direction. Biró worked on the gearbox with an engineer friend for more than a year until they had a design they were happy with. They signed a contract with General Motors guaranteeing them each $100 a month for the next five years (around £1,025 today), although the design never went into production.

Biró's next job was as a journalist for the weekly newspaper *Előre*. Visiting the print room of the newspaper one day, he became frustrated as the heat of the machinery caused his Pelikan fountain pen to leak. Watching the rotary printing machines, he began to wonder if a similar mechanism could be used to develop a new type of pen. One story claims that Biró was sitting in a café in 1936, trying to sketch out the details of how the mechanism would work. He was faced with a problem: a cylinder (like those used in the printing press) could only roll in one direction, but a pen needs to be able to roll in all directions. As he sat there, he watched some children playing with marbles in the street. One marble rolled through a puddle and he saw the wet trail the marble left behind on the ground; 'The solution flashed across my brain like a bolt of lightning: a ball!'

Biró was not the first to come up with the idea of using a ball to deliver ink to the page. John Loud of Massachusetts patented a pen with a metal ball for a writing point which was 'especially useful, among other purposes, for marking on rough surfaces – such as wood, coarse wrapping paper and other articles'. This was followed by a series of similar designs registered by other designers, although many of these early ballpoints had fairly broad writing points (two notable exceptions were the design from LaForest and Pignon who teamed up with Marcel Bich, and the Rolpen, marketed by Paul Eisner and Wenzel Klimes in 1935). Yet these pens were often unreliable, would leak or would simply stop working. Few worked as hard at refining their designs as Biró did, concentrating not only on the mechanism of the pen itself but also the ink inside.

László contacted his brother, who was now working as a dentist. As György had some knowledge of chemistry, he was put in charge of developing an ink suitable for this new pen. György visited a professor of applied chemistry and said that he was looking for an ink which 'remains fluid in the cartridge but dries as soon as it touches the paper'. The professor said what he was looking for was impossible. 'There are two kinds of dye,' the professor explained, 'those that dry quickly and those that dry slowly. What in heaven do you mean by a dye that makes up its own mind when it should dry fast and when it shouldn't? It does not, and cannot, exist.' For the next six years, Biró and his brother would try to prove the professor wrong.

Biró was still working part time at the newspaper, and still receiving monthly payments from General Motors, but despite these two revenue streams, producing prototypes was a costly business. A childhood friend, Irme Gellért, was able to offer some financial support, but Biró realised that in order to attract further investment, they would need to produce reliable demonstration models. Too often, the prototypes would leak, or wouldn't even work at all. Gellért and György had meetings with a series of potential investors, and as György talked business, Gellért would discreetly test the pens under the desk. If one worked, Gellért would present it to the investor. If it didn't, he'd pretend he didn't have a sample with him and arrange a follow-up meeting, promising to bring one along next time.

Biró and Gellért travelled to Yugoslavia to have a meeting with a banker named Guillermo Vig. As they arrived at the hotel ahead of their meeting, they used one of their demonstration pens to fill out the registration form at the reception desk (they managed to find one which worked). An elderly man standing next to them saw the pen and asked about it, introducing himself as General Justo and explaining that he was from Argentina and was interested in engineering. Biró and

Gellért went to the man's room to discuss the details of the pen. Justo told them that he believed there could be a market for such a pen in Argentina and offered to help organise visas for them both if they were interested. They arranged a follow-up meeting at the Argentinian embassy in Paris to discuss the deal a few months later. The meeting with Vig went well, and they agreed a distribution deal for the Balkan regions, estimated at around 40,000 pens a year. After the meeting, they told Vig about their meeting with the Argentinian and showed him the man's business card. Vig explained to the two men that the elderly man they'd met was the former president of Argentina, General Agustín P. Justo who was in Yugoslavia to promote trade links between the two countries.

Aware of the rise of anti-Semitism in the country, Biró was determined to leave Hungary before the end of the year. On 31 December 1938, he travelled to France. But with no real promising leads in France (and with his visa due to expire) he travelled to Argentina in 1940, with his brother following a year later. Biró went into business with Luis Lang, the husband of one of his brother's dental patients, to form Biró SRL; its first pen (the Eterpen) was produced in 1942. But the ink was still a problem; it would harden and the mechanism would fail, causing customers to return the pen as it became faulty. The small amount of money the company had quickly ran out. Lang turned to his solicitor, who introduced him to a man named Henry George Martin.

Martin was born in London in 1899 but moved to Argentina in 1924. Lang showed Martin the pen and he was clearly impressed. As the representative for a team of investors, he bought 51 per cent of the company. Martin played an important role in licensing the rights for Biró's designs to a newly formed partnership between Eversharp and Eberhard Faber in the US. He also formed a partnership with Frederick Miles of Miles Aircraft Ltd to create the Miles-Martin Pen Company in London in 1944. Returning to Argentina, Martin was approached by

American businessman Milton Reynolds who had heard about the ballpoint and was keen to acquire US rights for the pen. Martin's deal with Eversharp/Eberhard Faber had just been signed and so, unable to license the rights to Biró's design, Reynolds became determined to produce a pen of his own before Eversharp/Eberhard Faber were able to launch their ballpoint in America.

On 29 October 1945, Reynolds released the Reynolds International pen – the first ballpoint to be sold in the US. Reynolds signed an exclusive distribution agreement with Gimbels department store in New York, and publicised the launch of 'the miracle pen that will revolutionize writing' in the *New York Times*. The advert claimed that 'the fantastic, atomic-era, miraculous fountain pen that you've read about, wondered about and waited for' wouldn't need to be refilled for two years. (Reynolds's description of his product as a 'fountain pen' may seem confusing here, but at the time the phrase was used to refer to any type of pen which contained a supply of ink within its barrel – it was only once newer types of pens, such as the ballpoint, came along that a more specific terminology was required.) The pen also came with a guarantee:

> If the Reynolds International Pen fails to write during two years after the day of purchase, return it to Gimbels and we will immediately give you a refund.

Despite the high retail price of $12.50 (around $160 today), beating Eversharp/Eberhard Faber to market clearly paid off. According to the *New Yorker*, on the morning of the launch 'five thousand people were waiting to swarm through the doors, and fifty extra policemen were hastily dispatched to restrain the throng'. Ten thousand pens were sold in the first day; within three months, over a million had been sold. Reynolds would later say that the timing of that initial launch was crucial to its success. 'I knew the pen had to be selling by Christmas of 1945 to be a success,' he explained. 'The time had to be just right.

The public wanted a post-war wonder and it wanted it then. If the ballpoint pen had hit the market one year later, I don't think it would have sold worth a damn.'

Biró had spent years perfecting his design, which used capillary action as well as gravity to draw ink to the nib. The Reynolds pen, rushed into production, just used gravity. As a result, Reynolds needed to develop a different type of ink which would flow more easily. Reynolds called his new ink 'Satinflo', but it would smear on the page, would fade when exposed to sunlight, and soaked into the paper. Reynolds also failed to include an air-vent in the barrel. As the ink was used, a vacuum was formed, meaning the ink ceased to flow. The lack of an air-vent also meant that the pen would leak when it became warm (if kept in a jacket pocket, for example). Basically, the pen didn't work. In the first eight months, Reynolds replaced 104,643 faulty pens. Kenneth Parker, who at that time had not yet entered the ballpoint market, described the Reynolds International as 'the only pen that will make eight carbons and no original'.

In 1946, Eversharp finally launched its own ballpoint, the Eversharp CA. Based on Biró's patented capillary action design (hence 'CA'), the pen was a distinct improvement on Reynolds's rushed International ballpoint. The pen was premiered at a cocktail party at the St Regis hotel in New York. To demonstrate how strong it was, one Eversharp CA was hammered into a wooden block, while another was sealed in a pressurised jar to show it wouldn't leak if taken on a plane and then dropped into liquid nitrogen to show it could cope with extreme temperatures. As a spoiler, Reynolds introduced a new pen of his own (the Reynolds 400) on the same day that the Eversharp CA was launched at Macy's. Despite introducing a new model, Reynolds's technical problems continued – a point highlighted by Macy's sales staff who deliberately wore white gloves when demonstrating the new clean pen from Eversharp, unlike the staff at Gimbel's selling the messy Reynolds pens who couldn't do the same.

After the initial novelty of the ballpoint wore off, American consumers were left with a slightly leaky product that didn't really work. Although the Eversharp CA was a clear improvement on the Reynolds International, it still had its problems, and the market had been flooded with cheap imitations with exaggerated marketing claims. By the end of the 1940s, sales of ballpoint pens in the US had plummeted to around 50,000 a year. The bubble had burst, and it left an inky stain.

However, businessman Patrick Frawley was convinced the ballpoint still had potential. With an investment of $40,000, he bought the Todd Pen Company, changing its name to the Frawley Corporation. Frawley used a newly developed ink for his first Paper-Mate pen, launched in 1949. The following year, the company applied for a patent for its push-button mechanism, and the 'tu-tone' Retractable pen (priced at $1.69) restored consumer confidence in the ballpoint concept. A heavy advertising campaign in 1953, featuring celebrities such as Gracie Allen ('Gracie Allen says: "I simply adore Paper-Mate's style and smart new colors!"') and George Burns ('George Burns says: "Paper-Mate's push-button action is really great – always works!"'), plus a simple logo featuring two hearts next to one another, helped establish the Paper-Mate brand. In 1950, the company's sales were $500,000. By the following year, they had reached $2m, and by 1953, the company was turning over $20m in sales. In 1955, Frawley sold Paper-Mate to the Gillette Company for $15.5m – not a bad return on his initial investment of $40,000. Thanks to Frawley, ballpoints seemed respectable again.

Having concentrated on the fountain pen market (particularly following the success of the 51), Parker had been reluctant to launch a ballpoint until they were convinced they could produce a pen which wouldn't damage their reputation for quality. The company had briefly flirted with the idea, producing a novelty ballpoint based on the popular *Hopalong Cassidy* character in 1950, but despite the fact the pen was 'made and distributed by the Parker Pen Company', the company insisted it was 'not a Parker ballpoint pen' and that Parker was 'not in the ballpoint pen business'. Eventually though, the company succumbed and entered the ballpoint pen business in spectacular fashion. For several years, the design team at Parker worked on developing a retractable ballpoint before, in the autumn of 1953, launching 'Operation Scramble', giving themselves 'ninety days of hectic rush' to get a ballpoint pen from the drawing board into production.

The Parker Jotter was launched in January 1954 and is still on sale today. Developed by a research team of sixty-six people, the pen could write for six times longer than its competitors, was available in three line thicknesses (fine, medium and broad) and, crucially, didn't leak. Each time the push-button mechanism was pressed to extend or retract the nib, the writing point would rotate by 90° to prevent uneven wear on the housing. Despite all the effort put into ensuring the quality of the Jotter, Parker were still nervous of damaging their reputation, and the early pens did not feature the famous 'arrow' clip found on Parker's other pens; Kenneth Parker hoped that if the pen was a failure, people wouldn't associate it too closely with the brand. But he didn't need to worry – the pen was successful enough to earn its arrow within four years and, to date, over 750 million of the pens have been sold.

While the reputation of the ballpoint had been severely damaged in the US by opportunists like Milton Reynolds, it fared slightly better in Europe. Henry Martin's association with László Biró, coupled with Frederick Miles's background in engineering, meant that when, in 1945, the Miles-Martin Pen Company launched its first ballpoint in the UK (under the 'Biro' brand name) it was well received. A shortage of raw materials following the Second World War meant that production was limited, with demand outstripping supply, and so retailers were strictly limited to twenty-five pens each a month. Adverts for the pens claimed (with a quaint but crucial qualifier) that the

Biro would write 'for six months or more without refilling, according to the amount you write'. The pen was refillable, but customers would have to take it back to the shop for the new cartridge to be installed (some retailers offered a postal refill service – customers would send their empty pens away by post, then the refilled pen would be sent back to them the next day). By 1947, other British ballpoint pen manufacturers began to emerge, and within two years, there were over fifty such companies in the UK.

In order to remain ahead of their competitors, and to capture that all-important Christmas market, in 1949, Miles-Martin launched a range of novelty products. These included the Biro-quill ('This attractive Christmas novelty in six bright colours is a genuine quill with a Biro refill firmly cemented into the shaft. Biro-quill is an inexpensive gift which adds a touch of gay colour to any room') and, aimed at adults, the Biro Balita, a ballpoint pen with a cigarette lighter built in ('The only Christmas gift of its kind in the world').

In 1952, the company became Biro-Swan Ltd after buying Mabie Todd & Co. Ltd (manufacturer of Swan fountain pens). The same year, Henry Martin began legal action against Marcel Bich for infringement of László Biró's patents. Ironically, Bich's associate Jean LaForest had previously taken legal action against the Miles-Martin Pen Company, claiming their product infringed his 1932 patent. LaForest was unsuccessful in his bid and, again, the courts found in Martin's favour. The initial judgment stated that Bich should have all of his stock confiscated. Instead, Bich agreed a royalty deal with Martin, paying Biro Swan 6 per cent of the retail price of all pens sold, and 10 per cent on refill units. This agreement would remain in place until 1957 when *la société BIC* acquired 47 per cent of the shares in Biro Swan Ltd. BIC would go on to buy the rest of the company a decade later. The two companies were brought even closer together when John Martin (Henry Martin's son) married Marcel Bich's daughter, Caroline, in 1964. But while the Martin and Bich families were finally united by marriage and lived happily ever after, the Birós did not have such a happy ending. Biró had gradually been forced to sign away his shares in his own company to solve financial problems and to bring his family to join him in Argentina. Towards the end of his life, he worked as an advisor to the Argentinian pen manufacturer Sylvapen.

By the mid-1950s, ballpoints were outselling fountain pens by about three to one in the United States. A major factor in the ballpoint's favour was that fountain pens still had to be filled from bottles, which was both cumbersome and messy. In order to compete with the convenience and reliability of the ballpoint, Waterman launched the Waterman CF (cartridge filled) in 1954, the first pen to use a plastic ink cartridge. Ink cartridges had originally been introduced in 1890 by the Eagle Pencil Company, but the fragile glass cartridges would often break. Waterman had experimented with glass cartridges, producing models in 1927 and 1936, but faced the same problem. It wasn't until improvements in plastic manufacturing made plastic cartridges possible that the concept took off.

Despite the advantages offered by the plastic ink cartridge, the fountain pen will never be as cheap or convenient as the ballpoint. For some, however, this is part of its appeal. In 2012, the BBC reported that fountain pen sales were on the increase, with Amazon announcing that they had sold twice as many fountain pens as they had done in the previous year, and Parker celebrating the 'resurgence' of the fountain pen. It seemed to be making a comeback. The truth was, the fountain pen had already made several comebacks: 1980 saw a 'tremendous resurgence of the fountain pen's popularity'; in 1986, there was 'a resurgence of interest in expensive fountain pens on the part of affluent shoppers'; in 1989, the fountain pen 'emerged from the shadows' in which it had lain for years and was 'once again becoming a treasured fixture of daily life'; in 1992, it enjoyed a 'resurgence in popularity'; there was 'renewed interest' in the fountain pen in 1993; 1998 saw a 'major comeback of the luxury fountain pen'; in 2001, there was another 'resurgence of interest in the classic fountain pen'.

Part of the reason that this story of the fountain pen's perpetual comeback makes so many comebacks itself is because it manages to be just counter-intuitive enough to capture our attention without raising suspicions that there might not be much truth to it. With the rise of email, it's surprising that fountain pen sales are doing anything other than falling year by year. For them to remain stable (let alone make occasional upward spikes) seems odd. Yet there will always be a place for handwriting and as that place becomes smaller, we will cherish it even more. As Gordon Scott, Parker's vice-president for office products at Parker Pens in Europe, the Middle East and Asia, said during the fountain pen's most recent comeback, 'The relationship we have with a fountain pen is changing from it being a working tool towards more of an accessory.' In a world of email and iPhones, even a cheap fountain pen can be a status symbol – sending a message not necessarily about wealth, but about taste and refinement. Although, of course, if you want to show off how wealthy you are, you can do that too.

During a trip to the United States in 1906, German businessman Alfred Nehemias and engineer August Eberstein were impressed by the newly developed fountain pens they saw on sale there. Returning to Germany, they contacted Hamburg stationer Claus-Johannes Voss and decided to produce a pen of their own. Within a couple of years, the Simplo Filler Pen Company launched its first Rouge et Noir fountain pen. As if to symbolise the height of quality to which the company aspired, the next range of pens was named after Mont Blanc, the highest mountain in Europe. In 1913, the company adopted a six-pointed white star intended to represent the mountain's summit. The company would later adopt the mountain's name.

In 1924, Montblanc launched its first Meisterstück

('masterpiece') fountain pen, and in 1952 introduced the Meisterstück 149. The 149's black resin body and lid featuring three gold-plated rings has barely changed since its introduction. Each nib is engraved with the numbers 4810, a reference to the height (in metres) of the real Mont Blanc.

Under the Meisterstück brand, Montblanc began to sell a range of luxury items; not just fountain pens, but also watches, leather goods and jewellery. Montblanc introduced the precious metal Meisterstück Solitaire collection in 1983, which included the solid gold Meisterstück Solitaire 149. In 1994, the diamond-encrusted Montblanc Meisterstück Solitaire Royal (featuring 4,810 diamonds) entered the *Guinness World Records* book as the world's most expensive pen, priced at £75,000. This was later dwarfed by the $730,000 Montblanc Mystery Masterpiece (made in conjunction with Van Cleef) produced in 2007. $730,000 for a pen. What a world.

But of course, for many, such luxury is simply unimaginable, and the cheap appeal of the ballpoint is irresistible. With ballpoints becoming increasingly popular throughout the 1950s, the consumer magazine *Which?* put twenty-one of the leading pens to the test in 1958. 'The pens were tested to see whether they leaked,' explained the magazine, 'and how messy they got under different conditions; how well they wrote; how long the refill lasted; how well they were made.' In order to ensure each pen was exposed to precisely the same conditions, a series of elaborate tests was designed. The scientific approach.

Which? believed that 'a well-designed pen should not leak at heights', and so the pens were flown to a height of 15,000 feet, spent an hour at that altitude in an unpressurised cabin and then returned to ground level. This was repeated twelve times, with the conclusion that 'none of the pens showed any sign of leakage, either at the tip or inside the body'. So far, so good.

In order to find out 'what would happen if the pens were kept in an inside pocket', the pens were hung ('point downwards') in an oven at 90°F for twelve hours and then checked for leaks. They were then checked again twelve hours later. Finally, the experiment was repeated with the oven at 120°F – 'a temperature which might be reached if the pens were left in the sun, or in a handbag or jacket pocket on a radiator'. I can't help but feel that a simpler and perhaps even more accurate way to find out what happens if you leave a pen in a jacket pocket on a radiator would be to take a pen and leave it in a jacket pocket on a radiator. Sadly, the Queensway 100 and 125 leaked during this test, as did the Rolltip Rota Retractable and Rolltip Model '22'. These pens were rejected as the magazine did not consider that 'any pen which is liable to leak in a pocket or handbag is satisfactory'.

Which? did also try to find out which pens wrote the best – perhaps more important than how they coped when flown to 15,000 feet or put in an oven at 120°F. They did this by building an elaborate writing machine 'to eliminate the variable human element'. The machine held several pens and tested them 'simultaneously and under identical conditions'. Once placed in the machine, the pens traced out a shape 'somewhat like a capital D, of such size that, when 100 feet of paper had passed beneath the pens, each had traced out a line about a mile long' (the scientific precision of the experiment slightly undermined by the use of the word 'about').

Three of each refill (and in some cases six, to check results) were tested in the machine. 'This involved the examination of over a thousand feet of paper containing, in all, well over 130 miles of written line.' That must have been a fun day at the office. And rewarding, too, as the exercise revealed that 'none of the types was completely free from a tendency to blot' and that 'the refills of the same brand varied considerably'. The Platignum Kleenpoint Slim and the Scripto 250 were ultimately viewed to offer the best value for money 'as far as the essential qualities of a ballpoint pen are concerned'.

One pen which would be sure to pass all of *Which?* Magazine's tests is the Fisher Space Pen. There's a popular urban myth which contrasts the approach of the American and Russian space programmes. The myth-busting web site Snopes quotes an email which was widely circulated during the late 1990s:

> Thought for the day.
>
> During the space race back in the 1960s, NASA was faced with a major problem. The astronauts needed a pen that would write in the vacuum of space. NASA went to work. At a cost of $1.5 million, they developed the 'Astronaut Pen'. Some of you may remember it. It enjoyed minor success on the commercial market.
>
> The Russians were faced with the same dilemma. They used a pencil.

The story is intended to show the importance of 'thinking outside the box' and how often the simplest solution to a problem is the best. No doubt this was why Edward de Bono, often described as the 'father of lateral thinking', included the story in his 1999 book *New Thinking for the New Millennium*. However, the story is completely bogus.

In actual fact, the crew of *Gemini 3* (launched on 23 March 1965) did carry pencils on board. There had been some controversy at the time when it emerged that the thirty-four pencils had been purchased for the mission at a total cost of $4,382.50 (which works out at $128.84 per pencil – around $960 each today), although only two were ever actually taken on board. When asked to justify the cost of these pencils, NASA's Robert Gilruth explained that 'the actual writing mechanism was obtained from a small pencil procured from a local office supply house at a cost of $1.75 each'. However, the resultant cost of $128.84 per pencil included 'the fabrication and assembly of a take-up reel, baseplate, and the pencil housing'. You can't just buy a regular pencil from the store and take it up into space with you. You need a take-up reel, baseplate and pencil housing too. Obviously.

During the earlier Mercury missions, the crews had used grease pencils but these were considered 'unsatisfactory' for several reasons; 'cumbersome gloves made handling difficult, no good method of preventing "floating", possibility of getting away and jamming critical gear etc'. To prevent the risk of the pencil floating away, 'an initial effort was made to utilize a common pencil used by waitresses and other clerical people, which has a spring-loaded, retractable device'. This 'common pencil' would only cost $1.75 per unit. Unfortunately, 'it was found in testing this device that the spring clip was a gravity device and would not work in a weightless environment'. The design, fabrication and testing of an alternative to be produced in a limited quantity led to the $128.84 per unit price. 'If quantities of the type usually being procured for office or general supply use were required,' explained Gilruth, 'this unit price would be drastically reduced.'

In any case, in a zero gravity environment, a pencil is hardly ideal. The tip can easily break off and cause damage to sensitive equipment or drift into the eye of an astronaut. A pen would be much better. But just as it's untrue to say that NASA overlooked the simple solution of the pencil in favour of something more complicated, it's equally untrue to say that they then spent millions of dollars looking for that alternative.

The money spent on developing the Space Pen didn't come from NASA at all. The product was developed by inventor Paul C. Fisher, and was entirely self-funded. During the Second World War, Fisher had worked in a factory producing ball bearings used in aircraft propellers and this may, in part, have given him an appreciation for the precision engineering that went into manufacturing the tiny metal balls used as the writing points for ballpoint pens. Following the war, he developed a 'universal refill' ink cartridge that could be used in a variety of different ballpoint pens. Previously each manufacturer had sold their own refills, specifically designed to fit their own pens. 'The shortcomings of this situation and the inconvenience it causes the public and the pen dealers are obvious,' Fisher explained in his 1958 patent application. 'No one retailer is likely to stock refill cartridges of all the numerous manufacturers of ballpoint pens. Often a customer has to visit several retail establishments before finding a refill cartridge which fits his particular make and style of pen.' But as well as trying to solve the problems faced by ballpoint pen customers in the US, Fisher had bigger ambitions.

In 1960, Paul Fisher ran against John F. Kennedy in the New Hampshire presidential primary and hijacked one of JFK's rallies at the University of New Hampshire when he 'vaulted over the press table onto the stage'. Fisher demanded that he be given equal time to speak to the crowd – Kennedy agreed, commenting dismissively that 'The Constitution provides that the President must be American – an American born citizen – and 35 years of age. These qualifications Mr Fisher and I have in common.' Kennedy won.

After failing to become president, Fisher returned his attention to ballpoints. No doubt inspired by his rival's 1962 pledge to put a man on the moon before the end of the decade, he began developing an 'anti-gravity pen'. Fisher invested over a million dollars of his own money into developing a pen with a pressurised ink cartridge that 'writes in outer space' and could write at any angle 'including upside down'. Fisher sent his pen to NASA for testing and impressively, it passed all of their quality requirements. In fact, it wasn't the performance of the pen which NASA took issue with, but Fisher's marketing materials. Draft copy sent by Fisher for one advert claimed the pen had been 'developed by Paul Fisher for America's Space Program'; NASA suggested inserting the words 'possible use in' between 'for' and 'America's'.

The copy also claimed that the Fisher pen was 'the only pen that can write in the air-less, gravity-less conditions of outer space'. In fact fibre-tip pens had also been used in space and so it was suggested that this was amended to state that it was 'the only ballpoint pen' that could write in such conditions. Despite these difficulties, NASA did place an order for several hundred of Fisher's pens at between $4 and $6 dollars apiece for use on the Apollo missions. Fisher was then able to legitimately say that his pens had been used by US astronauts in space.

At that price, even at the height of the Space Race, the US astronaut market offered limited commercial potential. Fortunately for Fisher, the Space Pen also appealed to the billions of people who didn't have immediate plans to go into space. When Jerry sees Jack Klompus using the pen in the *Seinfeld* episode *The Pen*, he is clearly impressed and asks Jack about it. 'This pen? This is an astronaut pen. It writes upside down. They use this in space,' Jack explains. 'A lot of times, I write in bed,' says Jerry, 'and I have to turn and lean on my elbow to make the pen work.' As Jerry clearly recognised, a pen that can write upside down has advantages even if you aren't in space (although saying that, I'm not entirely sure if concepts such as 'upside down' actually exist in space), but Jerry shouldn't have taken the pen when Jack offered it to him. The Fisher pen can write 'wherever you need it – in the freezing cold of -45°C, in the blazing heat of +120°C, in the gravity-free vacuum of space, underwater, over grease – even upside down!' The use of the phrase 'even upside down' in that description bothers me. It suggests someone happily using the pen despite a potential temperature difference of 165°C, while suggesting that lying on your back and writing something in a notebook is some kind of extreme activity.

With the irresistible rise of the ballpoint throughout the 1950s, many people began to raise concerns about the effect this invention could have on that specific activity for which it was designed: handwriting. In the 1955 book *Teach Yourself... Handwriting*, John Le F. Dumpleton writes:

> [The ballpoint pen's] biggest drawback from the calligraphic point of view is its stylographic tip, which produces a line of uniform thickness. The easy glide of the ballpoint

> also prevents the pen from exerting any discipline over the writer. Nevertheless, in the hand of a practised penman it can achieve very satisfactory results for informal writing requirements.

The key, it would seem, is how you hold the pen. Parker would advise in the early instruction booklets for their famous ballpoint that 'better results and longer life' could be obtained by using the pen 'at a more upright angle than is usual with a fountain pen.' With a traditional fountain pen, the ideal angle between the pen and the paper is about 45°, but the housing of the ballpoint tip requires the pen to be held at an angle closer to the perpendicular. The uniform line thickness produced by the ballpoint was thought to remove individual character from handwriting. To combat this criticism, Biro Pens Ltd hired the services of 'handwriting expert' Frank Delino for the 1951 British Industries Fair:

> In eleven days, Delino interviewed 680 people and read their characters from writing done with a Biro. In nearly every case the 'patient' had to admit that the reading was entirely accurate. So much for the myth that Biro takes style and character out of writing.

Delino was a proponent of graphology, the analysis of handwriting to reveal personality traits. Newsreel footage from the time shows him identifying stage and film actress Sheila Sim as being 'artistic' and suggesting that Gracie Fields showed 'great determination' on the basis of their signatures. 'To Delino, it's all a science,' the voiceover explains, 'and a business.' Mainly a business.

Delino had a slight advantage as he studied the signatures of Sheila Sim and Gracie Fields: he would have already been aware of who they were and could tailor his analysis accordingly. Knowing who the subject is will obviously influence the interpretation given by the graphologist. During the 2005 Davos Economic Forum, a journalist from the *Daily Mirror* got hold of a sheet of notes and doodles and asked a graphologist what it revealed about Tony Blair. 'He is struggling to concentrate and his mind is going everywhere,' said Elaine Quigley of the British Institute of Graphologists, 'but he knows he will get to the bottom of the problems in time. That's Teflon Tony.' *The Times* quoted another graphologist who claimed the doodles showed Blair was 'an aggressive, unstable man who is feeling under enormous pressure'. A few days later, it emerged that in fact the doodles were by Bill Gates, not Tony Blair. 'We are astonished that no one who ran the story thought to ask No 10 if the doodles were in fact Mr Blair's, particularly as it was obvious to anyone the handwriting was totally different,' a Downing Street spokesperson said at the time.

So if graphology is a bogus pseudoscience (and it definitely is), is there nothing we can learn about someone's character based on an examination of a sample of their handwriting? Perhaps there are clues we could pick up on, which aren't based on the 'open ovals' or 'rising lines' of graphology but something more obvious: the colour of ink used. In Kingsley Amis's *Lucky Jim*, published in 1954, Jim Dixon receives a letter on a sheet of paper 'hastily torn from a pad, bearing a few ill-written lines in green ink'. The colour of ink seems to be intended to suggest the letter writer is of dubious character (L. S. Caton, who wrote the letter, is later in the book revealed to be a plagiarist), although as the sheet of paper had been 'hastily torn' from the pad, perhaps Caton was in a rush and it just so happened that the first pen he could find was green. The link between green ink and eccentricity is made a bit more explicit in Carl Sagan's 1973 book, *The Cosmic Connection*. Sagan describes a letter he received:

> There came in the post an eighty-five page handwritten letter, written in green ballpoint ink, from a gentleman in a mental hospital in Ottawa. He had read a report in a local newspaper that I had thought it possible that life exists on other planets; he wished to reassure me that I

was entirely correct in this supposition, as he knew from his own personal knowledge.

The phrase 'green ink brigade' soon emerged to describe people who write long, slightly confused letters to journalists and politicians often explaining some outlandish theory or revealing evidence of a conspiracy. If green ink is so closely associated with paranoid conspiracy theorists, it is fittingly ironic, then, that the most famous user of green ink was Mansfield Cumming, the first director of MI6. Cumming would sign his correspondence with his initial C in green ink – a tradition continued to this day by the current director, Sir John Sawers.

The range of ink colours used in ballpoints and fountain pens is limited by the dye-based inks used by these pens. Ballpoints use a viscous, oil-based ink – the thick, paste-like texture prevents the ink from flowing out of the air-vent when the pen is turned upside down but it is difficult to produce a multitude of colours in this form. Fountain pens use a thinner, more watery ink and would clog if used with a pigment-based ink, which contains solid particles of pigment suspended in a liquid medium.

Very early inks such as those used in ancient Egypt consisted of soot or ash mixed into a paste with water, gum or beeswax. Ochre would be used to create red inks. In China, soot was ground into a fine powder and mixed with animal glues to produce ink sticks. Water would be applied to the end of the ink stick, which was then ground against an ink stone to make liquid ink. In India, bones, tar and pitch were burnt to produce carbon black which would be mixed with water and shellac to produce a substance called 'masi'.

In AD 79, Pliny the Elder described the process of making black pigment in his *Naturalis Historia*:

> Black pigment can be produced in several ways from the soot obtained by burning resin or pitch, and this has led to the construction of factories which do not discharge their

> smoke into the atmosphere. The most highly rated black pigment is derived from resinous pine-wood.
>
> The final stage in the manufacture of all forms of black pigment is to expose it to sunlight. Black for ink is mixed with gum; black for walls, with glue.

Indigo would be used to produce blue colouration. From around the fifth century, iron gall ink became common. Produced by adding iron salts to tannic acid, the ink would initially appear quite pale on the page, but would then darken as it became fixed and permanent. This ink would be used up until the nineteenth century, however some formulations of the ink could be quite corrosive and would gradually attack the paper, causing it to disintegrate. The corrosive nature of iron gall ink meant that it was not suitable for use with fountain pens as it would attack the inside of the pen itself and so new ink formulations were developed.

In 1963 the Japanese pen company OHTO developed the rollerball; a ballpoint which used a water-based ink rather than the oil-based inks used by other manufacturers – its water-based ink made it easier to write with, as the thin ink would allow the writing ball to roll over the page and this smoothness gave the pen its name. Following this, many companies began to produce their own rollerball pens. The water-based ink meant that more colours could be produced using water-soluble dyes. Keen to develop a similar pen of their own, the Sakura Colour Products Corporation realised they were already behind their competitors and so, rather than launching another rollerball like everyone else, they developed a gel-based ink. The Sakura team researched 'thixotropic' materials – gels which are extremely viscous when at rest, but which become thinner and more liquid when disturbed. After several years of experimentation (using a wide range of materials including egg whites and grated yam) the team produced an ink which combined the properties of an oil-based ink and a water-based ink, and patented their discovery in 1982. The gel ink they created meant that solid pigments could be used rather than liquid dyes. Powdered aluminium and ground glass can also be added to the gel to create a metallic sheen or glittery sparkle. As a result, many more colours could be produced – the Sakura Gelly Roll range now includes seventy-four different colours and shades across their 'Moonlight', 'Stardust', 'Metallic' and 'Classic' collections.

So much choice, yet I still prefer the simplicity of black ink on white paper. It adds authority to my words, an authority I lack in all other areas of my life.

CHAPTER 3

Had a love affair but it was only paper

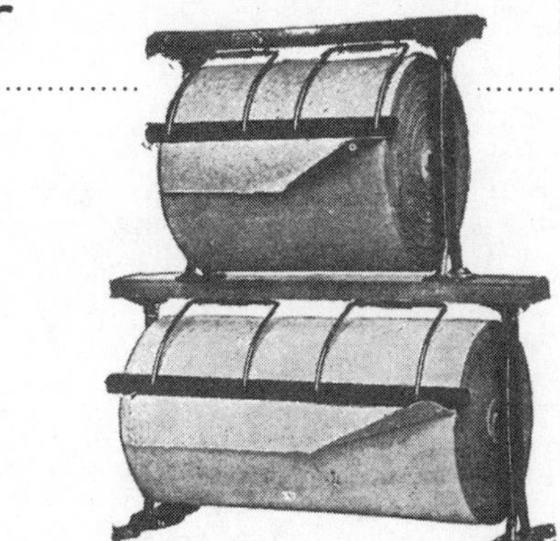

At the back of my desk, next to an orange and cream 8" × 5" index box (Velos 85, bought on eBay) is a stack of three little black notebooks. These contain scribbles, doodles, hurriedly scrawled ideas and reminders to myself. I keep them, although I'll probably never read through them again (and even if I were to read them again, I doubt I'd understand half of what's written). The notebooks are, of course, Moleskine. Without doubt, few notebooks are able to provoke such strong reactions as the Moleskine. For those who use it, this small black notebook inspires almost religious fervour. But it is also derided as a symbol of pretension; an ostentatious prop used to convey an image of creativity by its owner, sitting in a hipster coffee bar in any city the world over. Moleskine, MacBook, flat white.

Inside each Moleskine is a small booklet detailing the history of the notebook. The Moleskine is the 'heir and successor' to the 'legendary' notebook used by such notable figures as Vincent van Gogh, Pablo Picasso, Ernest Hemingway and Bruce Chatwin. The important words here are 'heir and successor'. The notebooks used by van Gogh, Picasso, Hemingway and

Chatwin weren't actually Moleskine notebooks, it's just that the notebooks they used were sort of similar to the ones Moleskine sell.

> A simple black rectangle with rounded corners, an elastic page-holder, and an internal expandable pocket: a nameless object with a spare perfection all its own.

The name itself comes from travel writer Bruce Chatwin. Writing in *The Songlines,* Chatwin describes his fondness for the '*carnets moleskine*' sold by a *papeterie* on Rue de l'Ancienne Comédie in Paris (the name 'moleskine' here referring to its 'black oilcloth binding'). 'The pages were squared,' Chatwin writes, 'and the end-papers held in place with an elastic band.'

In 1986, Chatwin was about to leave Paris for Australia and went to the *papeterie* to get a supply of these notebooks to last him during his trip. The shop's owner informed him that it was getting harder and harder to find a decent supply of the notebooks and the one supplier she did have was not answering her letters. When billionaire industrialist Howard Hughes heard in 1968 that Baskin Robbins were planning to discontinue his favourite flavour of ice cream (Banana-Nut), he placed an order for 1,500 litres of the stuff (a few days later, he switched to French Vanilla and it took the hotel which he owned and where he was living over a year to give the stuff away to their guests). Similarly, hearing his beloved notebooks were at risk of extinction, Chatwin decided to buy a hundred of them. 'A hundred will last me a lifetime,' he wrote. However, unlike Hughes, Chatwin was too late. Later that afternoon, Chatwin returned to the *papeterie* for news of his order. 'At five, I kept my appointment with Madame. The manufacturer had died. His heirs had sold the business. She removed her spectacles and, almost with an air of mourning, said "Le vrai moleskine n'est plus."' The real moleskine is no more.

Long before Chatwin, long before France even, humans had been writing things down on bits of paper. And before paper, there was papyrus. The papyrus plant, *Cyperus papyrus*, grows in shallow water and produces triangular woody stems about 4.5 metres tall and around 6 centimetres across. In around the third millennium BC, these stems began to be used to produce the writing material which takes its name from the plant.

In *Naturalis Historia*, Pliny the Elder explains how the Egyptians produced papyrus. The stem was split into thin leaves ('due care being taken that they should be as broad as possible'). These leaves were then placed on a surface which had been 'moistened with Nile water; a liquid which, when in a muddy state, has the peculiar qualities of glue'. The leaves are laid upon the table lengthways, 'the jagged edges being cut off at either end; after which a cross layer is placed over it. When this is done, the leaves are pressed close together, and then dried in the sun; after which they are united to one another, the best sheets always being taken first, and the inferior ones added afterwards.'

Papyrus remained the most important writing material for thousands of years until a dispute between King Eumenes II of Pergamon and Ptolemy V of Egypt sometime around 190 BC. The dispute resulted in the supply of papyrus to Pergamon being prohibited. Pliny writes that it was here that 'parchment was invented'. In fact, parchment had been produced for centuries before this, but it was during this period that the process to produce it was refined at Pergamon (parchment takes its name from the city via the Latin *pergamenum*). Parchment was produced from animal skin (calves, sheep or goats were most commonly used). The skin was removed from the animal and washed in a lime bath to remove the hair and the skin was then stretched on a wooden frame. Any remaining hair would be scraped away with a knife and the skin was then left to dry on the frame. Particularly fine parchment was made from calf skin and was known as vellum (from the Latin *vitulinum* for veal or calf).

Parchment and vellum had several advantages over papyrus.

Although relatively stable in the dry climate of Egypt, papyrus would become brittle in wetter conditions such as prevailed in western Europe. Unlike papyrus, which could only be written on one side, with parchment both sides could be used and it was much more durable – indeed it was possible to reuse parchment by scraping or washing away any writing on it (reused parchments being known as 'palimpsests' from the Ancient Greek *palímpsestos* meaning 'rubbed smooth again'). Parchment, and particularly vellum, also produced a much smoother writing surface than papyrus, allowing for a much more delicate writing style to emerge, particularly when used with the quill. Despite this, there was some resistance to its widespread introduction – the Roman physician Claudius Galenus was supposed to have complained that the shiny surface of parchment strained his eyes. Such delicacy – the poor man would no doubt have been blinded by the very pages you are reading right now. However, by the middle of the fourth century, parchment had begun to challenge the dominance of papyrus.

A decade after Chatwin's disappointment at the *papeterie*, the *moleskine* was reborn. Modo&Modo, a small publishing company based in Milan, decided to bring 'the legendary notebook' back to life. They selected this name with its 'literary pedigree' to 'revive an extraordinary tradition'. The first manufacturing run was for 5,000 pieces, and they were sold to stationery suppliers around Italy. Within a couple of years, they were supplying the notebooks throughout Europe and the US. Now the brand is global. In 2006, SGCapital Europe bought Modo&Modo for a reported £45m 'with the objective of fully developing the potential of the Moleskine brand' and in 2013, the company was valued at €430m when its shares went on sale on the stock exchange.

So now, this once humble 'nameless object' has become a global phenomenon, complete with the requisite set of branding guidelines and rules concerning the use of the MOLESKINE® trademark online. As the company explains on its web site, although they would be 'happy to allow' free use of their trademark 'to all those people who find in the name Moleskine a deep connection with their way of being and acting', they are unable to because its 'story and characteristics', as well as the 'tradition which has been accompanying it from many years', means they 'require a very careful evaluation of every single situation'. They ask that anyone wishing to use the 'name and/or trademark' online should request guidance from them directly ('Each request will be valued carefully by us and will receive a reply'). The trademark MOLESKINE® should not be used as a 'synonym of generic denominations such as, for example, "note pad", "block-notes", "agenda", etc', the company insists.

The enormous success of Moleskine as a brand, with the company's clever marketing – melding deft literary borrowings and vague historical associations – means it was inevitable that other similar products would soon come along hoping to share in their success. And this is why Moleskine seem so jumpy about people using their trademark to refer to these other 'generic denominations'. But if its supposed heritage is not enough, what else is there to make the Moleskine so special? The company would argue that it's the quality of their notebooks which sets them apart from the competition; 'there's a level of quality and fit and finish which has yet to be surpassed by any mass produced notebook', the company claims.

Recently, Moleskine has found its position as the preferred notebook of stationery aficionados in jeopardy with the UK launch of Leuchtturm1917. As the name of the Leuchtturm1917 notebook suggests, the company was founded in 1917, but their products weren't distributed in the UK until Carolynne Wyper and Tracy Schotness introduced them to the country in 2011.

Do the thread-bound pages and acid-free paper of the Moleskine and the 'ink-proof paper' of the Leuchtturm1917 justify their premium prices (particularly now high street stationers have also created their own notebooks based on the small-m moleskine ideal)? How sensitive are customers to details like this anyway? When SGCapital bought Moleskine, an interesting thing happened. They made a couple of small changes to the way the notebooks were packaged. A small note was added:

> Bound and Printed in China.

The sudden appearance of this statement of origin led some people to think Moleskine had moved production to China. In fact, from the very first batch, they had always been made in China, it just wasn't mentioned anywhere on the product. There wasn't any real difference between pre-SGCapital Moleskines and those produced after the takeover, but on Moleskine message boards and fan sites, people began to claim the 'new' Moleskines were inferior, in some loose, indefinable way:

> Ever since Moleskine moved their production to China ... every book is a little different than the previous one. The cover feels different, or the binding is tight, or it smells funny, or something.

Some, in an unconscious echo of Chatwin himself, even attempted to stockpile 'genuine' Moleskines. ('Soon after I discovered the Moleskines being made in China, I went to a nearby Borders and bought some that are not made in China to stock up.') Is the perceived quality of these notebooks just an example of the placebo effect? We think they're better than their rivals just because we think they're made by Italian artisans and when we discover they're mass-produced in China, we assume the worst? The irony is that people assume 'made in China' suggests poor quality, but China is essentially the birth place of paper and for centuries led the world when it came to paper production.

Cai Lun has often been credited with inventing paper as we know it today. Although the Egyptian method of producing papyrus created a useful material for writing and is the source from which we get the word 'paper' itself, there is a difference between the two materials. In the production of papyrus, layers from the plant are laid across one another and compressed to form a single sheet. Whereas paper is made from 'fibre that has been macerated until each individual filament is a separate unit'. This was how Cai made his paper.

Cai was a court eunuch, serving under Emperor He of the Dong Han dynasty. In AD 89, he was promoted and given responsibility for developing weaponry and military equipment. During the course of this work, he saw the importance of developing a cheap writing material. According to Cai's biography in the *Book of the Later Han* (written in the fifth century), 'in ancient times, writings and inscriptions were generally traced upon pieces of bamboo, or upon strips of silk which were given the name *chih*. But silk being costly and bamboo heavy, these two materials could not be used conveniently.' Lun 'conceived of the idea of making paper from the bark of trees, hemp waste, old rags and fish nets'. In AD 105, he presented his discovery to Emperor He 'and was highly commended upon his competency'. Cai was also rewarded financially and in AD 114 was made a marquis. Emperor He died in AD 105, and his wife assumed power. Following her death in AD 121, He's nephew, Emperor An, began to assert his authority and removed many of He's advisors from power. Facing a prison sentence, 'after bathing and dressing himself in his finest and most elaborate robes, he drank poison'.

Although for many centuries Cai Lun has been credited with the invention of paper, in 2006, a scrap of paper featuring written characters was found in Gansu in north-west China, which dates back to 8 BC, more than a hundred years before Cai's invention. Following the discovery, Fu Licheng from the Dunhuang Museum described the skill with which it was made as 'quite mature', showing that the material had been in use for some time. However, Mr Fu said that this should not be seen as diminishing Cai's achievement, noting that 'Cai Lun's contribution was to improve this skill systematically and scientifically, fixing a recipe for papermaking'.

Regardless of whether Cai had made the discovery himself, or had formalised a process that had already been known for at least a century, the production of paper following him greatly benefited Chinese culture. Paper was used not only for the transmission of ideas through writing, but for many other purposes; the decorative arts, business administration and credit, household furnishings and for sanitary purposes (toilet paper became common around the sixth century). The precise details of Cai's papermaking recipe are not known, but the process would have involved boiling the bark or cloth to soften it. Water would then be added and the mixture would be beaten either with mallets or in a pestle and mortar to produce a pulp. The pulp would be spread across a screen, allowing the water to drain away and producing a mat of fibres which would be placed on to a board to dry. The paper would finally be polished with a stone to produce a smooth writing surface. As the paper manufacturing process became more sophisticated, the screen would be dipped into a vat of pulp rather than having the mix spread across the screen by hand, but in principle, the process would remain unchanged for centuries.

As China began to build trade links with the Arab world, knowledge of paper began to spread (the Arabic word for paper, *kāghid*, is thought to derive from the Chinese *ku-chih* meaning paper made from the paper mulberry tree). Following the defeat of the Chinese by Arab armies at the Battle of Talas in AD 751, it is believed that two Chinese papermakers were seized as prisoners of war and forced to reveal their secrets in exchange for freedom. This may not be entirely accurate, but regardless, shortly after the Battle of Talas, paper production began in Samarkand. In 794, a second paper mill was built in Baghdad. By the ninth century, paper production had spread to Damascus and Tripoli. From there, it slowly expanded across the Arab world. In the tenth century, Fez became a centre of paper production and it is thought that it was from here that it was introduced to Europe, with the first paper mill being built in Xátiva in Spain in around 1150. And so the idea that cheap Chinese manufacturing ruined the quality of a product produced by an Italian company's interpretation of a British writer's recollection of the notebooks he bought in France seems a little patronising to say the least.

My Moleskine notebooks contain notes and scribbled thoughts, some entirely incomprehensible now. The early pages seem hesitant; the handwriting nervous and too considered. A new notebook can be so daunting. It takes a while to loosen up and get into it, to accept that it's OK to cross things out and make mistakes. Could this be a downside of Moleskine's premium prices? If a Moleskine notebook costs more, you should use it for something special to justify the money spent. The Bodleian library at Oxford University has kept the original moleskine notebooks used by Bruce Chatwin. The contents of them show that while some include notes for his various books or *Sunday Times* articles, they also include 'notes by Elizabeth Chatwin of shopping lists, chores and recipes'. If the Chatwins can use their small-m moleskines for shopping lists, I can use my big-M Moleskines for the same. I should relax.

If the cost of a premium Moleskine gives me performance anxiety, perhaps I should look to the other end of the price range. Less intimidating than the austere black Moleskine is the cheerful orange Silvine Memo Book. These narrow, flimsy notebooks are the complete opposite of everything the Moleskine represents – they're cheap and disposable, stapled together rather than stitch-bound and can be found in any newsagent or convenience store. And yet, unlike the carefully crafted, elaborate (though largely mythical) history of the Moleskine, Silvine is a brand with a genuine heritage.

William Sinclair was born in Otley, Yorkshire in 1816. After completing an apprenticeship as a printer and bookbinder with

William Walker in 1837, Sinclair opened his own business in nearby Wetherby but returned to Otley in 1854. The town of Otley had a thriving printing industry, and this was boosted by the 1858 invention of the Wharfedale printing press (one of the earliest cylinder printing presses). After Sinclair died in 1865, the business was continued by his two sons (the almost identically named Jonathan and John) and the company has now been run by six generations of the Sinclair family. In 1901, the company registered the Silvine trademark, and today the company produces more than three hundred products under this brand. The orange Memo Book was introduced to the Silvine range in the 1920s, along with the cash books and exercise books which also continue to be produced today.

The cheap disposable notebooks such as those produced by Sinclair wouldn't have been possible without the advances in paper manufacture that took place in the nineteenth century. Since Cai Lun's time, there had been very little development in the paper manufacturing process. Water-powered mills had been introduced in the thirteenth century, which removed much of the effort involved in turning the raw materials (usually hemp cloth or rags) into pulp. However, paper was still made in individual sheets on screens dipped into the pulp. This was time-consuming work – in his 1855 book on *Paper and Papermaking*, Richard Herring (not the comedian) describes how a sheet of 'Antiquarian' paper (the name used for a sheet 53 inches by 31) was made:

> So great is the weight of liquid pulp employed in the formation of a single sheet, that no fewer than nine men are required, besides additional assistance in raising the mould out of the vat by means of pulleys.

The physically demanding work of paper production, with its associated labour costs, was not something which any self-respecting industrialist could be expected to accept, and so attempts were made to mechanise the process. In 1790, French engineer Louis-Nicolas Robert began working at the Didot paper mill in Essonnes, France. He was frustrated by the demands of the members of the papermakers' guild and began to design a machine to reduce the reliance on mill-workers.

After a few months, he presented his plans to mill-owner Pierre-François Didot who described them rather bluntly as 'feeble', but nonetheless encouraged Robert to continue with his idea. Robert built a model of his machine as a prototype, but when tested, the machine didn't work. Despite the setback, Didot still had faith in Robert, but felt his energy could be better spent elsewhere. He moved Robert to a different part of the business; for the next six months, he worked on secondment in the flour mill. After imposing this break on Robert's experiments, Didot suggested that he should try again and provided him with a small team of technicians. The new team developed a working prototype ('not larger than a bird organ'). Satisfied with this new prototype, they built a larger machine capable of producing paper twenty-four inches wide (suitable for creating sheets of the popular Colombier size paper). Robert showed two sheets of this paper to Pierre-François's son, Saint-Léger, who was so impressed that the next day he arranged for the pair of them to travel to Paris to register a patent for the new invention.

Where previously, individual sheets of paper had been made on wireframes, Robert's machine consisted of a continuous wire mesh loop. A rotating cylinder poured the liquid pulp on to the wire mesh, and as the mesh loop was drawn forward, water would drain out of the pulp into a vat underneath. The pulp would then pass under a felt-covered roller which would squeeze out any remaining water, producing a sheet of paper as wide as the machine but of an 'extraordinary length'. 'The work of operating the machine can be done by children,' wrote Robert.

Despite the encouragement shown to him by the Didots, Robert would eventually fall out with the family. He sold his patent to them for 25,000 old francs to be paid in instalments (around £40,000 today), but when the Didots fell behind in their payments in 1801, he took back his patent. Following the French Revolution, progress with the invention was slow. In 1799, Saint-Léger Didot wrote to his brother-in-law John Gamble to see if there was potential for developing the machine in England. Gamble contacted Henry and Sealy Fourdrinier, a pair of stationers from London who showed interest in the machine and, with the help of a mechanic named Bryan Donkin, they built a machine based on Robert's designs. Over the next six years, the Fourdriniers spent £60,000 (around £5.8m today) on developing the machine with Donkin, but a problem with their patent application meant the design was quickly copied without the Fourdriniers receiving any reward for their hard work.

Today, modern papermaking machines are still based on the principles of the Fourdrinier machine, although with one important difference: the method of drying and smoothing the paper. On the original Fourdrinier machines, once the water had been squeezed out of the paper, each sheet would still need to be cut and hung to dry. On a modern machine, the paper passes over a series of heated drying cylinders while still on the roll. The dried paper then passes through a pair of pressure rollers to smooth it and ensure it is of uniform thickness.

During the development of the Fourdrinier machine, the pulp used to make paper was still produced from rags and scraps of cloth. But it quickly became difficult to find enough rags to keep up with demand for paper. By the middle of the nineteenth century, Britain was using 120,000 tons of rags a year to produce paper it relied on. Three-quarters of the rags and cloth used to produce this paper had to be imported (mainly from Italy and Germany). An alternative material was needed.

In 1801, Matthias Koops published his spectacularly titled *Historical Account of the Substances Which Have Been Used to Describe Events, and to Convey Ideas from the Earliest Date to the Invention*

Of Paper. While most of the books at the time were printed on paper made from the traditional rag cloth pulp, Koops's book was made from straw. However, the final few pages were printed on a different material; in the appendix, Koops states:

> The following lines are printed upon Paper made from Wood alone, the produce of this country, without any intermixture of rags, waste paper, bark, straw, or any other vegetable substance from which Paper might be, or has hitherto been manufactured; and of this the most ample testimony can be given, if necessary.

Koops also registered a patent in the same year for his method of 'manufacturing paper from straw, hay, thistles, waste & refuse of hemp & flax, & different kinds of wood and bark, fit for printing & other useful purposes'. Koops's method involved cutting the wood into shavings, and soaking the shavings in lime water, adding soda crystals and then boiling. This mixture would then be washed and boiled again, with the excess water being squeezed out before being made into paper 'by the usual & well-known processes of making paper'. 'In some cases,' Koops notes, 'it has been found to be advantageous to suffer the pressed material to ferment & heat for several days before reducing it to pulp, in order to its being made or manufactured into paper.' Although he included wood within his list of ingredients for producing paper, the discovery that wood could be used for this purpose wasn't made by Matthias Koops. It wasn't even made by humans. It was made by wasps.

In 1719, the French scientist René Antoine Ferchault de Réaumur noticed that the material used by wasps to build their nests was very similar to paper:

> The American wasps form a very fine paper, like ours; they extract the fibres of common wood of the countries where they live. They teach us that paper can be made from the fibres of plants without the use of rags and linen, and seem to invite us to try whether we cannot make fine and good

> paper from the use of certain woods. If we had woods similar to those used by the American wasps for their paper, we could make the whitest paper, for this material is very white. By a further beating and breaking of the fibres that the wasps make and using the thin paste that comes from them, a very fine paper may be composed.

Réaumur also commented that 'the rags from which we make our paper are not an economical material and every paper-maker knows that this substance is becoming rare'. Despite this observation, Réaumur never developed the idea.

The idea was picked up by a few who experimented with using different materials, but it was left to Koops to really develop the concept. Unfortunately for him, the experiment would prove to be a costly one. Koops and his investors spent around £45,000 (£2.8m today) building an enormous paper factory in the area which would later become known as Millbank in London, but despite the factory successfully producing paper from straw and other materials, it was not enough to recoup the costs of the investment and the company soon went bankrupt. In December 1802, workers at the factory wrote to the shareholders informing them that following the closure of the mill, 'a great number of wet packs of paper are spoiling in the vat-house and the pulp is rotting in the vats'. They went on to say, 'We should be happy to be informed of your determination as our discharge is our misfortune and not of our own seeking.' Considering the way in which they had been treated, the letter was signed, with admirable restraint, 'Your very obedient, though distressed servants'. It was several decades before anyone would be able to find a successful method of producing paper from wood; then after those decades of waiting, two came along at once.

Charles Fenerty was born in Nova Scotia in 1821. His family owned a lumber yard in the Canadian woods and from an early age, Fenerty worked in the saw mills assisting the workers logging the nearby forests. Throughout the 1820s and 1830s, a number of paper mills began to be built throughout Canada, but supplies of rags were running low. Fenerty began to experiment with using wood fibre to form paper in the early 1840s. In 1844, he wrote to his local newspaper, enclosing a sample of his new paper:

> Enclosed is a small piece of PAPER, the result of an experiment I have made, in order to ascertain if that useful article might not be manufactured from WOOD. The result has proved that opinion to be correct, for – by the sample which I have sent you, Gentlemen – you will perceive the feasibility of it. The enclosed, which is as firm in its texture as white, and to all appearance as durable as the common wrapping paper made from hemp, Cotton, or ordinary materials of manufacture is ACTUALLY COMPOSED OF SPRUCE WOOD, reduced to a pulp, and subjected to the same treatment as paper is in course of being made.

As Fenerty was only in his early twenties at the time of his invention, he struggled to have his idea accepted by the Canadian papermakers, who considered him some young upstart. But around the same time, a German weaver, Friedrich Gottlob Keller, was granted a patent for his wood-grinding machine. By forcing blocks of wood against a wet grindstone, Keller broke the wood down to its fibre and was able to produce wood pulp. The first paper he produced contained 40 per cent cloth fibre for strength, but later experiments would produce entirely wood-based paper. In 1846, Keller sold his patent to Saxony-based papermaker Heinrich Voelter. Voelter went into business with engineer Johann Matthäus Voith, and the pair of them began to mass-produce the machines Keller had designed, ultimately leaving Keller with nothing.

Wood paper soon replaced rag-fibre paper in almost all situations, although one area where wood paper is still not used is bank note production. The paper used for bank notes issued by the Bank of England is 'manufactured from cotton fibre and linen rag, which makes it tougher and more durable than the more common wood pulp paper'. The paper is supplied by a specialist paper manufacturer, who literally have a licence to print money. In 2013, the Bank of England announced they would begin to print bank notes on polymer (beginning with a new £5 note featuring Sir Winston Churchill to be introduced in 2016), which is 'cleaner, more secure and more durable than paper'. The lifespan of a polymer bank note is around two and a half times that of its paper rival. As well as durability, some believe that the notes boast an additional feature. When the Bank of Canada introduced a new polymer $100 note in 2013, rumours began to spread that it contained a 'scratch and sniff' panel, releasing the scent of maple syrup. Although many Canadians were convinced the notes smelt like the country's famous syrup, the bank denied the story. 'The bank has not added any scent to the new bank notes,' a spokesman told ABC News. Dr Marilyn Jones-Gotman from the Department of Neurology and Neurosurgery at McGill University in Montreal attributed the rumour to a widespread 'olfactory illusion' causing people to believe they can smell something that isn't there.

Once Keller's work had shown that wood pulp could be used to make paper, there soon followed a demand for a more efficient method of producing the pulp. Keller's method of mechanically grinding down blocks of wood produced paper that was fragile and would tend to yellow over time. This was because Keller's wood pulp contained lignin, a chemical compound found in tree cell walls. Once the compound had been identified, with its paper-weakening powers, it was obvious that it needed to be eliminated. Chemical processes were soon developed to remove lignin and therefore produce a stronger, more resilient paper (although the yellowing of old newspaper pages shows that the cheaper mechanical method was not replaced entirely).

One attempt at creating a purer paper came in 1851; Hugh Burgess and Charles Watt from Hertfordshire began boiling wood chips in caustic alkali under high pressure in the production of their wood pulp (known as the 'soda process'). This produced a whiter paper than had previously been possible, but they struggled to profit from their idea. Benjamin C. Tilghman, an inventor from Philadelphia, developed a process using sulphurous acid in the 1860s, but financial problems limited his progress. Europeans Carl Daniel Ekman and George Fry were the first to profit from the sulphite process. Following on from the work of Tilghman, Ekman used a solution of bisulphate and magnesia to produce wood pulp in 1872. Ekman had been born in Sweden, but moved to England where he teamed up with George Fry. In 1874, a mill was built in Sweden that used the Ekman-Fry sulphite method, and this method would remain the dominant process for manufacturing wood pulp until the 1940s when it was replaced by the kraft process. Discovered by German chemist Carl F. Dahl, this method used sodium sulphate to produce a stronger pulp and takes its name from the German word meaning power or strength.

Chemical processing of wood pulp and mechanical production meant that it was possible to produce paper of reliable quality and thickness both cheaply and quickly. This led to the introduction of standard sizes and thicknesses for paper products. There had already been several attempts to standardise paper sizes throughout history – perhaps the earliest is illustrated on the limestone tablet on display at the Archaeological Civic Museum of Bologna. The tablet dates from 1389, and the inscription on it reads:

> These are to be the molds of the city of Bologna, which say what the sizes of the sheets of cotton paper must be, which are made in Bologna and the surrounding area, as is set out here below.

Below the inscription is a nest of four rectangles illustrating the four sizes '*inperialle*', '*realle*', '*meçane*' and '*reçute*'. The terms Imperial ('*inperialle*') and Royal ('*realle*') would continue to be used (with slight variations in their associated dimensions) right through until the introduction of decimalisation, and it is even possible that the Imperial size is derived from the sizes used for ancient papyrus scripts.

Throughout the late Middle Ages, papermakers began imprinting watermarks on their paper as a guarantee of quality. These would be created by pressing a piece of shaped wire into the paper stock as it sat in the mould. Watermarks would serve several purposes; they were used to identify the papermaker, but also became associated with different paper sizes, such as 'foolscap'– the term used to describe a sheet of paper 13.5" × 17". The name for this paper size derives from the 'fool's cap' watermark ('The original figure has the cap and bells, of which we so often read in old plays and histories'). The foolscap watermark was introduced in the middle of the fifteenth century and although in Britain it was eventually replaced by the Britannia or lion figure, the name remained – and it's the only traditional paper size to be name-checked in a Brian Eno song as far as I know.

In 1786, the German physicist Georg Christoph Lichtenberg wrote to fellow scientist Johann Beckmann describing an exercise he had given to his students. Lichtenberg had challenged his students to find a sheet of paper which when halved still had the same length/width ratio as the original. In order to illustrate the point, he decided to enclose a sheet of such paper with his letter. 'Having found that ratio,' explained Lichtenberg, 'I wanted to apply it to an available sheet of ordinary writing paper with scissors, but found with pleasure, that it already had it. It is the paper on which I write this letter.' It had been intended as a difficult mathematical puzzle, but in fact, the answer had been there all along: the medium was the message.

Lichtenberg believed that this ratio had 'something pleasant and distinguished' about it and asked, 'Are these rules given to the paper makers or has this form spread through tradition?

Where does this form come from, which appears not to have emerged by accident?' In fact, the puzzle set by Lichtenberg for his students wasn't of his own devising. It had previously been set for (and solved by) Dorothea Schlözer in 1787. Schlözer, a German scholar and one of five members of the elite *Universitätsmamsellen* group of female academics, had apparently solved the problem (her own teacher had failed to answer the same question when he had originally been presented with it in 1755).

The semi-magical ratio which satisfies Lichtenberg's puzzle is 1:√2 (approximately 1:1.41). Taking a piece of paper made according to this ratio and cutting it in half (parallel to its shorter sides) creates two pieces of paper with the exact same aspect ratio. This is the principle used in the modern A-series (cutting a piece of A3 in half creates two sheets of A4), but appears to date back at least a thousand years (two of the four paper sizes illustrated on the Bologna tablet have a ratio of 1:142, conforming almost exactly to this rule).

A more comprehensive system was proposed by Dr Walter Porstmann in Germany following the end of the First World War. Porstmann was born in 1886 in Geyersdorf and studied maths and physics at university. Porstmann's first paper on standardisation was published in 1917, and this drew the attention of Waldemar Hellmich, director of the newly formed *Normenausschuß der deutschen Industrie* (Standardisation Committee of German Industry). Over the next few years, Porstmann developed his system and this was eventually published as a German national standard by the *Deutsches Institut für Normung* committee in 1922 as DIN 476. The standard was adopted by Belgium in 1924, and soon spread around the world. By 1960, twenty-five countries were using the system and by 1975 it was so common that it was published as international standard ISO 216 by the International Organisation for Standardisation.

The A-series specified in ISO 216 is based on the metric

system; the largest sheet in the A-series, A0, has a surface area of $1m^2$ and the dimensions 841 mm × 1189 mm (producing the Lichtenberg ratio of 1:√2). Halving each sheet parallel to its shortest side then produces the next sized sheet in the series:

Sheet size	*Dimensions (mm)*
A0	841 × 1189
A1	594 × 841
A2	420 × 594
A3	297 × 420
A4	210 × 297
A5	148 × 210
A6	105 × 148
A7	74 × 105
A8	52 × 74
A9	37 × 52
A10	26 × 37

ISO 216 also specifies the dimensions of the related B-series of paper sizes (mainly used in print design). A further ISO standard (ISO 269) was introduced in 1976 specifying the dimensions of the C-series used for envelope sizes – a sheet of A4 will fit unfolded into a C4 envelope and so on.

Until the reformation of the postal service in 1840, envelopes were not commonly used at all when sending letters through the post. Postage prices were based on the number of sheets each letter contained, and an envelope was considered a sheet and so would therefore be charged accordingly. To avoid this extra cost, letter sheets were used; these were single sheets of paper that could be folded and sealed without the need for an envelope. In 1837, the social reformer Rowland Hill published a pamphlet outlining his proposed changes to the postal service (*Post Office Reform: Its Importance and Practicability*). Hill suggested a simplification of the whole system, with dramatically reduced prices based on weight rather than the number of sheets used per letter. Most importantly of all, Hill proposed a uniform charge regardless of where in the country the letter was to be delivered, and with payment in advance (previously the costs would often be paid by the recipient of the letter). In 1840, Hill's proposals were accepted and two different methods of pre-payment were introduced – adhesive postage stamps, and pre-stamped letter sheets and envelopes.

The artist William Mulready was commissioned to design the artwork for these letter sheets and envelopes. Hill had expected these to be more popular than postage stamps but within days the overly elaborate designs with their images of Britannia with a lion at her feet, began to be caricatured and mocked. Hill wrote in his journal of his fear that there may be the need to 'substitute some other stamp for that design by Mulready' as 'the public have shown their disregard and even distaste for beauty'. The Mulready designs were quickly withdrawn and replaced with embossed envelopes. The envelopes featured an image of Queen Victoria based on a design by William Wyon. The Wyon penny envelope was an immediate success, but still could not compete with the convenience and flexibility of the postage stamp.

With postage costs reduced following Hill's reformations and the failure of the Mulready designs, demand for cheap mass-produced envelopes quickly grew (around 26 million letters were posted annually before the reformations – by 1850, this had increased to 347 million, 300 million of which were posted in envelopes). Previously, envelopes had been made by cutting diamond-shaped 'blanks' from rectangular sheets of paper, but this resulted in considerable waste. George Wilson, a stationer from London, was granted a patent for his design offering 'improvements in the cutting of paper for the manufacture of envelopes, and for other purposes' in 1844, which reduced the amount of waste by tessellating the envelope blanks. The following year, Rowland Hill's brother, Edwin, and paper manufacturer Warren de la Rue were granted a patent for their machine which not only cut envelope blanks, but also folded them into shape (before this, the blanks would be folded by hand around a tin template).

Different envelope types and shapes gradually evolved, each suited to a different function: 'pocket' envelopes, which are sealed along the shorter edge and 'wallet' envelopes, which are sealed on their longer side; the traditional diamond-shaped 'baronial'; the more rectangular 'booklet'; the 'ticket'; the 'side seam'; the 'announcement'. So many envelopes, so little time to explore them all. The traditional 'baronial' envelope, produced from a diamond-shaped piece of paper, had the advantage that it could be sealed at a single point, but the introduction of gum arabic for use on 'self-sealing' envelopes soon meant that this was no longer a concern – as one etiquette writer noted in 1855, although wax seals are elegant, 'they are less indispensable than they were before the invention of "adhesive" or self-sealing envelopes' (the same writer also warned against using incorrectly sized envelopes, saying 'Be careful not to put too large or thick sheets into small envelopes; it is in bad taste, and is, like crowding a fat hand into a small glove, a clumsy, awkward proceeding'). The gum arabic used on envelopes would need to be moistened, either by licking or through the use of a roller damper. It's not a fun job.

In the *Seinfeld* episode *The Invitations*, George manages to convince his fiancée Susan to buy the cheapest wedding invitations in the store, despite the fact that the store assistant tells him that particular type hasn't been manufactured for a number of years because of a problem with the glue. Luckily for George, the store still had a couple of boxes in storage. He leaves Susan to lick the envelopes ('Ugh, awful,' she says, disgusted by the taste of the glue). After licking the envelopes, Susan passes out and is taken to hospital. After breaking the news to George that Susan has died, the doctor asks if she had been exposed to 'any kind of inexpensive glue', explaining that 'traces of a certain toxic adhesive, commonly found in very low-priced envelopes' had been found in her blood. George answers that she had been sending out their wedding invitations, but justifies the cheap envelopes because they were expecting about two hundred people, which is understandable, I suppose.

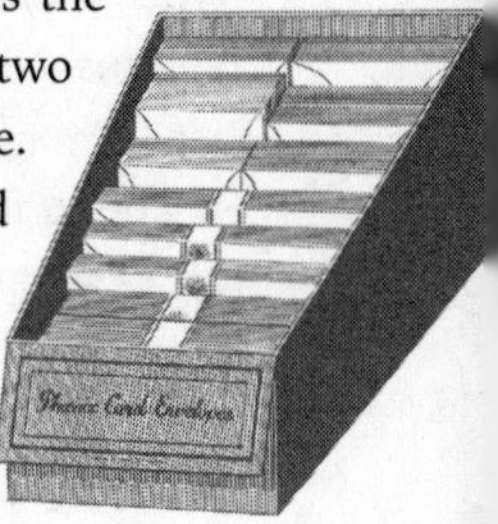

But is such a thing possible? Is envelope glue bad for you? In 2000, a story began circulating online:

> If you lick your envelopes ... You won't anymore!!!!
>
> A woman was working in a post office in California, one day she licked the envelopes and postage stamps instead of using a sponge. That very day the lady cut her tongue on the envelope. A week later, she noticed an abnormal swelling of her tongue. She went to the doctor, and they found nothing wrong. Her tongue was not sore or anything. A couple of days later, her tongue started to swell more, and it began to get really sore, so sore, that she could not eat. She went back to the hospital, and demanded something be done. The doctor, took an X-ray of her tongue, and noticed a lump. He prepared her for minor surgery.
>
> When the doctor cut her tongue open, a live roach crawled out. There were roach eggs on the seal of the envelope. The egg was able to hatch inside of her tongue, because of her saliva. It was warm and moist.
>
> This is a true story reported on CNN.

The story was comprehensively debunked by the Snopes web site, and yet, like all the best urban myths, the rumour still persists.

But while this particular story may not be true, that is not to suggest that envelopes are entirely harmless. In 1895, the *New York Times* reported on the death of a Mr S. Fechheimer, who died 'from blood poisoning as a result of cutting his tongue while licking an envelope'. Envelopes can kill, but these things don't happen often – so don't have nightmares and do sleep well.

Despite the simplicity of the ISO 216 paper sizing system, and despite the fact that it has been adopted by nearly every single country in the world, the US continues to resist its charms. No doubt this is due to the fact that the metric system has never been fully accepted in the country – whether this is due to fear ('The metric system is the tool of the devil! My car gets forty rods to the hogshead and that's the way I likes it,' as Grampa Simpson once cried) or plain inertia is hard to tell, but this reluctance does make the country something of an anomaly, as the *CIA World Factbook* explains:

> At this time, only three countries – Burma, Liberia, and the US – have not adopted the International System of Units (SI, or metric system) as their official system of weights and measures. Although use of the metric system has been sanctioned by law in the US since 1866, it has been slow in displacing the American adaptation of the British Imperial System known as the US Customary System. The US is the only industrialized nation that does not mainly use the metric system in its commercial and standards activities, but there is increasing acceptance in science, medicine, government, and many sectors of industry.

Catch up, guys.

The reluctance of the US to adopt the clearly superior ISO 216 system (and metrication generally) means that the country continues to use inch-based sizing formats such as 'legal' (8.5" × 14") and 'tabloid' (11" × 17"), although the most commonly used size is the 'letter' format (8.5" × 11"). The US letter format has obscure (and indeed, European) roots, being based on a paper frame introduced by Dutch papermakers in the 1600s. It was found that a frame of about 17" by 44" was pretty much at the limit of what any individual worker was capable of manipulating. A quarter of a sheet produced the US letter size.

As its name suggests, the larger 8.5" × 14" size is often associated with the legal profession. In 1884, a paper mill worker from Massachusetts named Thomas Holley began collecting scrap pieces of paper and binding them together to form cheap writing pads. Not restricted by the legacy of the physical limitations of Dutch papermakers some two hundred or so years earlier, Holley was free to choose his own page size for his notepads, and celebrated this freedom by adding three inches of extra length to each sheet. Holley formed the American Pad & Paper Company (AMPAD) to sell the pads. He was contacted by William Bockmiller, a stationery salesman with a special request. One of Bockmiller's customers was a judge who bought plain pads and ruled lines in them by hand. The judge wanted to buy a ruled version of the pad, including a margin so he could annotate his notes. Holley began producing ruled pads, and the legal pad was born. Traditionally, legal pads have always been yellow in colour, although the reasons for this are unclear. One theory is that as the original pads were formed using scraps from different paper mills, the pads were dyed yellow to give them a uniform appearance. Another theory is that using yellow sheets means the paper is easier to identify in a file of otherwise plain white paper. Regardless, the colouration has stuck, with some claiming that the yellow hue is easier on the eye than stark white.

Given Holley's role in the development of the legal pad, it's a little ironic that in November 1903, newspaper stories began to report financial irregularities within the company:

> It was discovered on Saturday that Thomas W. Holley, for a long time president and treasurer of the corporation, has misappropriated its funds. It is impossible just now to state just what the shortage is, but it appears to be about $35,000.

It was believed that Holley issued fraudulent stock certificates for the company and pocketed the proceeds. A few days later, the *Lowell Sun* reported that he was thought to have fled to Canada:

> Friends of T. W. Holley, the missing treasurer of the American Pad & Paper Company, intimate that he is in

> Canada, with a special interest in what course is taken toward getting warrants for his arrest.
>
> It is said all of his life insurance policies include a suicide clause, whatever that may be worth to him or his family now. He carried $30,000 insurance which it is stated was placed in such a manner as to provide for his family in event of the trouble that he foresaw must come sooner or later.

While Holley pioneered the familiar US yellow legal pad, he was far from the first to sell ruled writing pads. In 1770, John Tetlow from London was the first to be granted a patent for his 'machine for ruling paper for music and other purposes'. Seven years later, Joseph Fisher patented his 'Universal Machine' capable of producing 'lines for writing and drawing' which could also produce squared paper. Although for most of us, a machine capable of producing both lined and squared paper would more than adequately serve our needs, for some, the simple choice between plain, ruled or squared is not enough. For them, they reach for the Writersblok dot grid, or the Cuaderno series of notebooks designed by Spanish graphic designer Jaime Narváez. The Cuaderno series comprised:

> A collection of four notebooks in which, from the traditional lines and squares, new patterns are proposed. They can be seen as drawings, even though they maintain their functionality as a conventional notebook. An invitation to imagine and experiment new forms of drawing or writing, returning our gaze/attention to the medium itself.

Legal pads, such as those developed by Holley, are often staple-bound along the top of the page, with a perforation allowing each page to be torn out and filed. But there are many binding options: the stapled Silvine Memo book; the stitched Moleskine; the casebound Black n' Red; the spiral-bound jotter. I find jotter pads confusing. The spiral binding along the top makes it easy to take notes, flipping each page over at a feverish rate as words are quickly scribbled down, but I gradually lose sense of what is 'forwards' and what is 'backwards' within the notebook. I get lost by the middle of the pad. Normally, when trying to find a particular note or idea I hurriedly wrote down, I can at least remember if it is on the left or right page of the notebook or pad and then try to find it that way. The jotter pad offers only vague locative information. That crucial scribble could be anywhere. Notebooks need a 'search' function.

The growing popularity of note-taking apps like Notes on the iPhone and Evernote seems to offer one way of overcoming this problem. Allowing you to sync notes and thoughts between different devices in an easily searchable format threatens to make the habit of writing something down in a book made out of paper seem like an increasingly archaic habit. However, the two formats don't have to live in opposition. Moleskine's Evernote Smart Notebook is an attempt to combine the tactile pleasure of physical objects with the advantages of cloud computing and search. After writing in the notebook, you can tag notes with 'Smart Stickers' according to theme ('Home', 'Work', 'Travel', 'Action', 'Approved' and 'Rejected') and then take a photo of the page using the Evernote Page Camera. Depending on how good your handwriting is, the handwritten text should be searchable. I suspect it would struggle with my handwriting, but then again, I struggle with my handwriting. As we become increasingly reliant on screens, our typing skills may improve, but possibly at the cost of the legibility of our handwriting.

But there is an obvious solution. If we want the machines to be able to read our handwriting more easily, we need to practise more. Write by hand more often – your computer needs you.

They can't move until I pick up a pencil

Because we are introduced to them at an early age, before we graduate to the grown-up world of ink, it's tempting to believe that historically, the pencil must precede the pen. Its humble wooden body reinforces its simple nature; no plastic barrel or metal casing, just traditional earthly materials. But the wooden pencil is a more recent invention than one might otherwise assume. In fact, it's fair to say that the pencil is older than the 'pencil', though younger than the 'pencil'. While that last sentence might sound like gibberish, it has an element of truth. The word itself existed long before the lead pencil, and the lead pencil existed before the 'traditional' wood-encased pencil.

Deriving from the same Latin base as the word 'penis' (meaning 'tail'), the *penicillum* was a fine-pointed artist's brush used for intricate script work and drawing (the hairs of the brush being taken from animal tails, and so thankfully its seemingly crude origin actually has an innocent explanation and our blushes are spared). From here, it mutates into the Old French *pincel* and finally the Middle English word *pencil*. But it isn't until the early sixteenth century that it began to be used in the modern sense of a 'lead pencil' rather than a brush. The 'lead' here is also a misleading term – while lead had been used to produce writing styluses used by the Greeks and Romans, the modern lead pencil contains no lead, which is a relief as it is highly poisonous and not really suitable for use by primary school age children (unless they're children you don't like).

The pencil was born in the early sixteenth century following a stormy night in Cumberland. The exact year is lost to history, but the popular legend is that a particularly fierce gale uprooted a large oak tree in a field in Borrowdale near Keswick one night, exposing a deposit of a mysterious black substance: graphite. The new material was given the name *plumbago* because of its similarity to lead, although it was also called 'black lead, kellow or killow, wad or wadt, which words properly mean black'. It is from this similarity between graphite and lead that we have the idea of the 'lead pencil' – the persistence of this term being one of those odd linguistic quirks, in the same way that you might 'film' something on your phone or buy baked beans in an aluminium 'tin'.

Farmers nearby found that this substance – which they referred to as 'wadd' – was useful for marking their sheep and it was soon found to have other useful applications too. Sticks of wadd or graphite would be wrapped in string so they could be used without dirtying the fingers. By the 1560s, the lead pencil was known throughout Europe. In 1565, the Swiss naturalist Conrad Gesner published a book about fossils entitled *De omni rerum fossilium genere, gemmis, lapidibus metallis, et huiusmedi libri aliquet, plerique nunc primum editi* ('Of all the kind of things, fossils, gems, stones, metals, and the like; several books, most of them now published for the first time'). In this, he included an illustration of a pencil with the description:

> The stylus shown below is made for writing, from a sort of lead (which I have heard some call English antimony), shaved to a point and inserted into a wooden handle.

From this description, we see that the pencil Gesner is describing is not like the common wooden pencils we use today. The pencil does not have a lead core running all the way through it; rather Gesner's pencil (a version of which was produced by German pen manufacturer Cleo Skribent in 2006) features a wooden shaft into which the lead is inserted – in this sense, it has more in common with a *porte-crayon* or leadholder. The *porte-crayon* was an instrument used by artists to hold pieces of graphite, chalk or charcoal for sketching. Typically made of brass, the device consisted of a thin tube, which was split at both ends. The chalk or graphite was inserted into one end and held in place with a metal ring; a second colour could be inserted into the other end. The *porte-crayon* became increasingly popular throughout the seventeenth and eighteenth centuries, and is still used by some artists today (the hands emerging from the page in M. C. Escher's 1948 lithograph *Drawing Hands* are drawing each other with *porte-crayons*). While the *porte-crayon* was used to hold a variety of drawing materials (chalk, charcoal and graphite sticks), the leadholder was more specialised, as its name suggests. Featuring a wooden or metal body, sharpened pieces of graphite were inserted into a tapered metal holder and held in place by a ring with a screw thread. The leadholder was preferred by draughtsmen because of the precision of its line and it gradually evolved into the mechanical pencil.

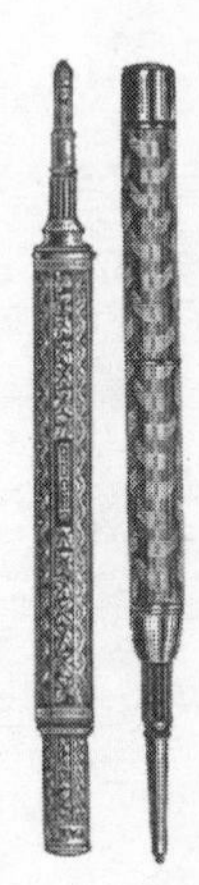

There is disagreement over when the modern pencil was born. Craftsmen in and around Keswick first began to encase graphite sticks in wood during the late sixteenth century (although some claim it was developed in Italy around the same time). Exporting blocks of graphite was discouraged and consequently the area quickly became the worldwide centre of pencil manufacturing. The Borrowdale mines were the only known source of high-quality pure graphite in the world, and so Cumbrian wadd quickly increased in value. The mines were closely guarded – at times they would even be flooded to prevent people stealing the valuable material. As it increased in value, the wadd would be transported by armed guard to London where it would be sold at auction for huge sums. The buyers were invariably Keswick pencil manufacturers and so the wadd would then be brought back up north, again by armed guard, to be turned into pencils. The pencils in Keswick were produced by cutting blocks of graphite into thin sheets. A groove would be cut into a narrow square rod of wood and the sheet of graphite would then be inserted into the groove where it would be scored with a sharp blade and broken off level with the wood. A thin piece of wood was glued over the top, encasing the graphite, and the rod would then be shaped into pencils.

Having developed as a global trading centre in the Middle Ages, the German city of Nuremberg was closely associated with the mining industry as traders from the city began to exploit mineral deposits throughout central Europe. Workers from Nuremberg were brought in to mine the Borrowdale wadd, sparking the interest of German merchants. These merchants began to manufacture their own pencils in Nuremberg using inferior local graphite mixed with sulphur and other binding materials. But the quality of these German pencils could not compete with the pure graphite pencils from Keswick.

When France declared war on Great Britain in 1793, an economic blockade between the two countries meant that the French no longer had access to pencils produced in Keswick. They couldn't even purchase the inferior German pencils, and so the French minister of war Lazare Carnot instructed Nicolas-Jacques Conté to develop a pencil that did not rely on imported materials. Conté had originally trained as a portrait painter, but following the French Revolution, turned his attention to science. It was said that Conté had 'every science in his head and every art in his hands'. Although French plumbago was not as pure as the Borrowdale wadd, Conté was familiar with the material and quickly developed a process for mixing powdered graphite with clay to produce thin rods that were then fired in a kiln. More brittle than pure graphite sticks, this new material was, however, superior to the German product. The principle of mixing graphite powder with clay, patented by Conté in 1795, is still the process by which pencils are produced today.

The Conté method meant that the world was no longer reliant on the Borrowdale mines, although the area of Keswick remained well known for producing the finest pencils in the world. Previously, the pencils had been made by hand in small workshops, but in 1832, the first pencil factory in the area was opened by Banks, Son & Co. This company would pass through several hands before becoming the Cumberland Pencil Company in 1916. Four years later, it was bought by British Pens Ltd, who built a new factory in Keswick, from which the company manufactured its pencils under the Derwent brand until 2008 when it moved to a new site in nearby Workington.

Adjacent to the site of the old pencil factory in Keswick is the Cumberland Pencil Museum. For £4.25 you can gain access to this charming (though small) museum, and visitors also receive a pamphlet and free pencil. The museum (which appears in Ben Wheatley's cult 2012 film *Sightseers*) features a variety of displays informing visitors about the history of pencil manufacturing in the local area, as well as a chance to see 'the world's longest colouring pencil'. Measuring 7.91 m (25 ft 11.5 in), the pencil was certified by Guinness World Records in May 2001 ('This is a REAL pencil, if you put a sheet of paper in front of the yellow pencil core, it would write yellow on the sheet of paper'). Weighing 446.36 kg (984.05 lb), it took twenty-eight men to carry it into the museum where it is now suspended from the ceiling.

As a result of Conté's discovery, the world no longer relied on pure graphite for its pencils. Thanks to its earlier association with the industry, Nuremberg began to establish itself as a rival to Keswick for the title of 'global centre of pencil manufacture' during the eighteenth century. The city is still home to two globally renowned pencil companies, Staedtler and Faber-Castell, although quite who came first remains a source of rivalry between them both. Three families of craftsmen had begun to produce pencils by hand in Nuremberg in the 1660s – the Jenigs, the Jägers and the Staedtlers. Indeed, the first person to be officially recognised as a 'pencil maker' was Friedrich Staedtler, stealing the honour from those early pioneers in Borrowdale. However, it was only the Staedtler family who continued to make pencils throughout the next few generations. Controlled by trade guilds, the city had strict rules about which new manufacturers could form companies within the city walls. As a result, it wasn't until 1835 when the restrictions were relaxed that Johann Sebastian Staedtler (Friedrich's great-great-grandson) was able to officially register the company J. S. Staedtler.

Just outside Nuremberg, in the small town of Stein, a cabinet-maker named Kaspar Faber began producing pencils in 1761. Faber had originally planned to set up business within Nuremberg but was put off by the city's trade restrictions and so moved further out. So while Staedtler has roots going further back in time by around seventy-five years, both companies laid claim to being the oldest. The dispute rumbled on until a court decision in the 1990s officially ruled in Faber-Castell's favour. Consequently in 2010, when Staedtler was celebrating its 175th anniversary, Faber-Castell was making plans for its 250th, even though Friedrich Staedtler had begun making pencils almost a hundred years before Kaspar Faber. Although I'm not sure how much difference the ruling really makes. I doubt there are many people who'd consider Staedtler to be some fly-by-night operation because they've only been around for 175 years.

Using Conté's mixture of clay and powdered graphite, European companies were able to compete with Keswick's finest. The new production method also allowed manufacturers to produce pencils in a variety of grades. By varying the ratio of clay to graphite, pencil leads could be made harder or softer (a hard pencil lead contains more clay than a soft pencil). Conté originally gave his pencils numbers to indicate how hard they were (with the number increasing as the pencil got harder). The London pencil company Brookman is believed to have introduced the 'H' and 'B' designation still used today ('H' meaning a hard pencil, which produces a thin line, and 'B' meaning a blacker or softer pencil). As the hardness of Brookman's pencils increased, so they increased the number of Hs, although as the manufacturing process became more sophisticated and more grades became possible, it made more sense to combine the numbering system with the 'H' and 'B' designations (it's easier to tell the difference between an 8H and a 9H than between a HHHHHHHH and a HHHHHHHHH, and a 9H is certainly easier to ask for in your local branch of Ryman). The most commonly used pencil sits as a happy medium between the worlds of H and B; the HB, a symbol of equality and harmony – we should all aspire to be more like the HB. In that one respect.

A similar system was developed in the US by the author Henry David Thoreau in the 1840s. Although better known today as an author and naturalist, his contribution to the history of pencil manufacturing is often overlooked. Thoreau spent a long time working at his father's pencil company in Concord, Massachusetts. John Thoreau's pencil company was already well regarded when his son joined the business, but Henry made an important contribution. It is not clear if he was aware of Conté's method of mixing graphite powder with clay or if he developed a similar process independently, but soon after beginning work at the company, Thoreau's pencils were available in four grades of hardness (in the US, this grading system still dominates, with the #2 roughly equivalent to the HB). John Thoreau had gone into business with his brother-in-law Charles Dunbar in 1823. Two years earlier, Dunbar had discovered a deposit of plumbago in New England but his lease on the new mine only lasted seven years and so he was determined to extract as much of the precious material as quickly as possible.

Dunbar was not the first American to manufacture wooden pencils, however. William Munroe, a cabinet-maker also from Concord, had begun to produce pencils in 1812. With the US at war with Britain, Munroe struggled to make a living selling the furniture he made. With pencils in scarce supply but in constant demand, Munroe reasoned, 'If I can but make lead pencils I shall have less fear of competition, and can accomplish something.' With no real scientific background, Munroe struggled for almost a decade before he was able to produce a pencil of decent quality.

An associate of Munroe's, Ebenezer Wood (has anybody got any Veras?), built a pencil mill by the Nashoaba Brook in Concord which featured a series of machines invented by Wood to increase efficiency. Wood's mill included a 'wedge glue press' which could hold twelve gross of pencils as the glue dried, a circular saw for cutting grooves in six pencils simultaneously, a saw for shaping pencil casings into hexagonal and octagonal shapes, and a machine to grind plumbago into powder. Munroe had a head start on Thoreau, but by the 1830s, the two were bitter rivals, with Thoreau's pencils much better quality. Trying to thwart the competition, Munroe attempted to convince Wood to refuse to grind Thoreau's plumbago. The gambit backfired; Wood was making more money from Thoreau than he was from Munroe and so he ended up cancelling Munroe's contract instead.

The pencil manufacturing world is a competitive place and just as there is debate over who can rightfully call themselves the oldest pencil manufacturer in Nuremberg, there are rival claims as to who built the first pencil factory in the United States. The inventor Joseph Dixon opened a graphite processing factory in Salem in 1827, and although it produced a small number of pencils from 1829 onwards, it seems that the first purpose-built pencil factory in America was created by the great-grandson of Kaspar Faber. Seeking a reliable source of cedar wood to produce pencil casings for the family business back in Nuremberg, Eberhard Faber moved to the US in 1849 as a representative of A. W. Faber, the company now owned by his father. Gradually, Eberhard recognised that the US had all the raw materials required to produce quality pencils and, in 1861, Eberhard opened his first factory in Manhattan.

Faber's biggest rival was born in 1799. Joseph Dixon's entrepreneurial mind was trained to capitalise on 'the opportunity offered by the suggestion of the moment', making him sound like a hopeful on *The Apprentice*. Much to Faber's relief, for a long time the opportunity suggested by the moment wasn't anything pencil related. Dixon operated a foundry producing a variety of products from local plumbago, using it to create polishes, lubricants and paints. As its name suggests, the Joseph Dixon Crucible Company was initially concerned with the production of heat-resistant graphite crucibles for use in industry, although Dixon gradually turned his attention to the world of the pencil. His first experiments in pencil manufacturing pretty much began and ended in 1829. They weren't good pencils. Instead, Dixon had much more success with his graphite crucibles used in the production of iron and steel. However, just like Conté and Munroe before him, Dixon benefited from war (War, what is it good for? Absolutely nothing – except improvements in pencil manufacturing methods, say it again).

The American Civil War created demand for cheap, reliable pencils so that troops could make quick notes in the field and send messages to each other. With his manufacturing background, Dixon quickly developed a method of mass-producing pencils of consistent quality – each pencil marked with his trademark crucible logo. By 1872, Dixon's company was producing 86,000 pencils a day and was the largest consumer of graphite in the world. In 1873, Dixon bought the American Graphite Company, based in Ticonderoga, New York – and it was this town that would lend its name to Dixon's most iconic product. Launched in 1913, the Dixon Ticonderoga wasn't the first pencil to be painted yellow, but it's certainly the most famous. Different colours had been used to coat pencils before, but perhaps because of their utilitarian existence, or simply because it matched the colour of graphite, most mass-produced pencils were painted black. The Koh-I-Noor 1500 pencil, introduced by the Czech pencil company Hardtmuth at the 1889 World Fair in Paris, broke from this tradition.

The Hardtmuth company had been founded almost a century earlier by Josef Hardtmuth, who opened an earthenware factory in Vienna in 1790. Within a decade, the company was producing pencils, using the Conté method of mixing graphite and clay. Taking inspiration from the Koh-I-Noor yellow diamond, Hardtmuth's yellow cased Koh-I-Noor 1500 was available in seventeen grades, unheard of at the time. Suddenly, yellow was the colour of quality, and Dixon, like many other American companies, adopted it as their default, hoping to capitalise on its positive associations.

The Joseph Dixon Crucible Company's response, the Dixon Ticonderoga, remains the most famous yellow #2 pencil in America. With its distinctive yellow and green ferrule (the metal sleeve holding the eraser in place), and produced in enormous quantities, the Dixon Ticonderoga became such a big seller that Joseph Dixon's company would eventually adopt its name.

But the Dixon Ticonderoga Company doesn't simply make pencils; it 'empowers people to take conscious and subliminal thoughts, facts, ideas and dreams, and preserve them using tools that are simply extensions of themselves'. Claimed (by Dixon Ticonderoga) to be 'the world's best pencil', the low cost and high quality of the pencil meant it quickly became a familiar sight in offices and classrooms across the US. But despite its popularity, the pencil does have a dark side – George Lucas apparently used a Dixon Ticonderoga when he was working on the first draft of the screenplay for *Star Wars Episode I: The*

Phantom Menace, and so the pencil is, at least in part, responsible for Jar Jar Binks.

Lucas isn't the only notable Dixon Ticonderoga user. It was the preferred pencil of Roald Dahl, who would sharpen six each morning before he could begin that day's writing. Dahl began using the Dixon Ticonderoga in 1946, after he became unhappy with the quality of the pencils in post-war Britain, which were like 'writing with a piece of charcoal on a lump of gravel'. It is not surprising that writers should be so sensitive to the qualities of different pencils given how dependent on them they are in order to work their craft. 'On the third finger of my right hand I have a great callus just from using a pencil for so many hours every day,' John Steinbeck once wrote. 'You see I hold a pencil for about six hours every day. This may seem strange but it is true. I am really a conditioned animal with a conditioned hand.'

Unlike the unforgiving pen or typewriter, the pencil accepts that writers make mistakes. And so before the word processor allowed entire paragraphs to disappear at the click of a button, it must have seemed like a less intimidating way to start a first draft than its more permanent, stricter alternatives. 'Everything I write for the first time is written with a pencil,' Nobel Prize winner Toni Morrison told the *Paris Review* in 1993, 'my preference is for yellow legal pads and a nice number two pencil.'

In his 1989 novel *The Dark Half*, Stephen King explicitly demonstrates the power that the pencil gives the writer over his creations. The book tells the story of Thad Beaumont, an author who publishes some of his novels under the pseudonym George Stark. Beaumont writes his novels on a typewriter, Stark's are written with a pencil. As Beaumont attempts to kill off his alter ego to focus on his other writing, the pencil begins to act as a totem, marking the difference between the two authors as the two do battle for their very existence. Stark eventually takes physical form and the pencil then becomes a weapon. The pencil Stark uses, referenced throughout the novel, is the Berol

Black Beauty. Stark's beloved Black Beauty began life as part of the Blaisdell Pencil Company range, although through a series of company buyouts and mergers, it became part of the Berol series. Stark's pencil of choice is no longer available – its closest relative being the Papermate Mirado Black Warrior, 'the world's smoothest writing pencil – guaranteed!'

Despite its apparent innocence – its temporality and the fact that we first use it as children before moving on to pens – there is undoubtedly a certain menace presented by the pencil. Whether it is the grim but thankfully apocryphal urban legend of the distraught student committing suicide using two pencils during a stressful exam, Thad Beaumont's eventual victory over his alter ego, or Heath Ledger as the Joker performing his 'magic trick' and making a pencil disappear in *The Dark Knight*, there's something about the pencil which suggests its suitability as a weapon. The wooden body and sharpened, shiny metallic tip resembles a miniature spear. Even if not used to cause physical harm, it can be a sign of mental distress – illustrated by Edmund Blackadder's attempt to show he has gone mad in *Blackadder Goes Forth* by sticking two pencils up his nose and putting a pair of underpants on his head; or the thumb-snap of James's Dixon Ticonderoga in *Twin Peaks* when he hears that Laura has been murdered (was it a stunt pencil or does actor James Marshall just have very strong thumbs? I've tried, and I can't snap a pencil like that). Perhaps it is because of its impermanence, that it can so easily be erased, that the pencil has this apparent association with menace and madness. As a writing medium, it never has to take responsibility for its own actions. It can act on impulse – something can be pencilled into your diary without you fully committing. There is always a way out.

Of course, most of the time, a character or alter ego doesn't physically manifest itself and attempt to kill an author. That happens very rarely. But still, the pencil can give life and can take it away from any fictional character. 'It would be a great joke on the people in my book if I just left them high and dry, waiting for me,' John Steinbeck noted in one letter to a friend. 'If they bully me and do what they choose, I have them over a barrel. They can't move until I pick up a pencil.' Fortunately for them and for his readers, he did pick up his pencil. In fact, during his career, he tried a number of different pencil brands in his search for 'the perfect pencil', while recognising how elusive such a thing would be:

> For years I have looked for the perfect pencil. I have found very good ones but never the perfect one. And all the time it was not the pencils but me. A pencil that is all right some days is no good another day. For example, yesterday, I used a special pencil soft and fine and it floated over the paper just wonderfully. So this morning I try the same kind. And they crack on me. Points break and all hell is let loose.

When Steinbeck found a pencil he liked, he would buy dozens of them at a time. He tried the Blaisdell Calculator and the Eberhard Faber Mongol 480 ('which is quite black and holds its point well'), but Steinbeck's favourite was the Blackwing 602:

> I have found a new kind of pencil – the best I have ever had. Of course it costs three times as much too but it is black and soft but doesn't break off. I think I will always use these. They are called Blackwings and they really glide over the paper.

The Blackwing 602 was launched by Eberhard Faber in 1934. Its unique soft lead was the result of adding wax to the graphite and clay mix, meaning it could write with 'half the pressure, twice the speed', according to the slogan printed on each pencil. The pencil also featured an unusual ferrule to hold its rectangular eraser. Normal ferrules are simply round metal sleeves attached to the end of the pencil into which the eraser is fixed. Once the eraser has been worn down, there isn't much that you can do. The flat ferrule of the Blackwing, however, features a metal clip that allowed the rectangular eraser to be removed and extended as it wore away.

Steinbeck was not the only fan of the Blackwing: the arranger Nelson Riddle, best known for his work with Frank Sinatra, considered the pencil his favourite; Quincy Jones always kept one in his pocket when he was working; Vladimir Nabokov name-checks the pencil in his final novel *Look at the Harlequins!* ('I caressed the facets of the Blackwing pencil you kept gently twirling'); animator Chuck Jones described his art as 'a flurry of drawings created by a Blackwing pencil'. But despite the Blackwing's celebrity fan base, the pencil was discontinued in 1998. Eberhard Faber had been bought by Sanford in 1994, and around the same time, the machine which made these unique clips and metal ferrules broke down. As the company still had a large stock of the ferrules, Sanford decided not to repair the machine. After four years, the stock of metal clips was finally exhausted and the pencil was no more. The decision not to repair or replace the machine was based on the fact that by the mid-1990s, the company was only producing around a thousand dozen Blackwings a year – in a factory capable of producing that volume in a single hour. Although not a problem for its celebrity admirers, the Blackwing's premium price (two to three times that of its rivals) severely limited its commercial appeal – particularly in a world where high volume stores like Staples and Office Depot were rapidly replacing independent stationery retailers and slashing prices to attract customers.

Since its death in 1998, the Eberhard Faber Blackwing 602 has obtained near mythical status. Articles in the *Boston Globe, Salon* and the *New Yorker* have celebrated the Blackwing and writer Sean Malone has spent the last three years documenting the history of the 602 on his blog *Blackwing Pages*. In 2005, seven years after they had been discontinued, the American composer Stephen Sondheim told one interviewer that he still uses Blackwings, having bought several boxes before they'd been discontinued. 'I sometimes get letters, "Do you have any source for the Blackwings?"', Sondheim added. Unsharpened 602s sell on eBay for between $30 and $40 for a single pencil, but the people buying them aren't just enthusiasts adding to their collections; they buy them to use them.

Aware of its reputation and discovering that the Blackwing trademark had recently expired, in 2010 Charles Berolzheimer of the California Cedar Products Company launched the Palomino Blackwing – a tribute to the Eberhard Faber original. Made with a soft lead and with a similar flat ferrule and removable eraser, the Palomino Blackwing was released to mostly positive reviews, although some hardcore enthusiasts complained that the new pencil did not feature the iconic HALF THE PRESSURE, TWICE THE SPEED slogan. The Palomino Blackwing 602, launched the following year, was even closer to the original and included the famous slogan.

However, some were not entirely enthusiastic about this development in the Blackwing story. In fact, one of the most outspoken critics of the Palomino pencil was Sean Malone, author of the *Blackwing Pages* blog. In much the same way as Modo&Modo appropriated the legends of Chatwin and Hemingway in their publicity for the Moleskine, Malone believes that California Cedar Products have similarly attempted to blur the history of the Eberhard Faber pencil with their modern tribute. The Palomino web site lists Chuck Jones, John Steinbeck and Leonard Bernstein as Blackwing users, and says that the pencil is 'claimed by many to be the best writing utensil in the world'. Malone has attacked the marketing claims of California Cedar Products in a series of posts on his blog, describing such sleight of hand as 'cultural vandalism – the wanton inclusion of such great names in their ads to try and sell their pencils, but without regard to the facts or to the legacies of those great people, or the actual cultural history of the Blackwing 602'. The lesson here seems to be clear: don't mess with a pencil enthusiast.

The obvious irony of Blackwing devotees paying huge

amounts of money on eBay for increasingly rare Eberhard Faber 602 originals is that as they use each pencil, they are killing the very thing they love. Each twist of the pencil sharpener shortens its life. As Edmund explains to Elizabeth I in an episode of *Blackadder II*, 'Madam, life without you was like a broken pencil – pointless.' The sharpener literally gives the pencil its point in life, but kills it at the same time. I know marriages like that.

The earliest wooden pencils would have been sharpened with a knife – just as a knife would be used to shape the point of a quill. However, by the nineteenth century, dedicated pencil sharpeners began to be produced. In 1828, Bernard Lassimone of Limoges in France was granted a patent for his *taille-crayon* or pencil sharpener. Robert Cooper and George Eckstein from London began selling the exciting sounding Styloxynon in 1837, which consisted of 'two sharp files neatly and firmly set together at right angles in a small block of rosewood' and could produce 'a point as fine as a needle'. By the middle of the century, small handheld sharpeners (such as those developed by Walter Foster of Maine in 1855) became common. The basic form has changed little since.

Given the attention paid to pencils by authors and writers, it's not surprising that they would be concerned about the best way to sharpen them too. The pencil offers a crude approximation of word count – a sign of progress more satisfying than simply counting pages. Hemingway believed that 'wearing down seven number-two pencils is a good day's work'. During his time in Paris, he would always carry a notebook, two pencils and a pencil sharpener ('a pocket knife was too wasteful'). In *A Moveable Feast*, he describes sitting in a café as he sharpened his pencil 'with a pencil sharpener with the shavings curling into the saucer under my drink'. The simple sharpener such as that used here by Hemingway is familiar to all of us. The pencil is inserted into the sharpener and rotated by hand, a blade slicing off a thin sliver of wood, like the peeling of an apple skin. However, during the second half of the nineteenth century, larger mechanical sharpeners would also start to be used. Often fixed to a wall or to the surface of a desk, the pencil would be held in place by gripping teeth and a handle would rotate the cylindrical blade, producing fine pencil shavings and a sharp, even point. This is the type of sharpener Nicholson Baker enthusiastically described to the *Paris Review*:

> The pencil sharpener was probably the best thing about school back then, actually – a little chrome invention under your control. It had a thundering sound, a throat-clearing sound, that I especially liked – Ticonderoga is almost onomatopoetic. And of course I oversharpened and broke the point, so I got to stand there for a while making that sound, ticonderoga ... oga ... oga.

The mechanical sharpener was quicker and produced a better point than the single-bladed handheld version, but more convenient still is the electric sharpener. Initially developed at the beginning of the twentieth century for use by pencil manufacturers rather than pencil users, electric pencil sharpeners gradually made their way into the office and home over the next few decades. Considerably more expensive than its more simple rivals, the electric pencil sharpener is aimed at the high-volume pencil user. John Steinbeck was a fan:

> The electric pencil sharpener may seem a needless expense and yet I have never had anything that I use more and was more help to me. To sharpen the number of pencils I use every day, I don't know how many but at least sixty, by a hand sharpener would not only take too long but would tire my hand out. I like to sharpen them all at once and then I never have to do it again that day.

'When in my normal writing position, the metal of the pencil eraser touches my hand, I retire that pencil,' Steinbeck wrote (the rejected pencils being given to his young children). Yet the rather obvious fact that pencils get shorter as they are sharpened was overlooked by the Bureau For At-Risk Youth in New York in 1998 when they handed out pencils to pupils of a nearby school printed with the slogan 'TOO COOL TO DO DRUGS'. As the students quickly realised, sharpening the pencil changed the message initially to 'COOL TO DO DRUGS' and then simply 'DO DRUGS'. It was only when a ten-year-old pupil pointed out the problem that anyone realised. The company then started printing the slogan in the opposite direction, so that it would say 'TOO COOL' once sharpened down instead. 'We're actually a little embarrassed that we didn't notice that sooner,' a spokeswoman later admitted.

Normally, a shortened pencil isn't too problematic – uncomfortable to hold perhaps, but nothing which is likely to lure schoolchildren into a life of narcotic depravity. Once a pencil gets too short to use comfortably, most people will switch to a new one – the old one being relegated to a life sitting in a little pot next to the phone, or in a drawer somewhere with some bits of string and batteries which may or may not still work. But the useful life of a pencil can be extended. Pencil extenders come in many forms, but at their most basic, they consist of a shaft with an opening at one end into which the pencil stub can be inserted. A screw mechanism or metal ring holds the pencil in place, and the stub can continue to be used right down to the bitter end. There is, however, a danger with some of the more elaborate pencil extenders that you may look like Burgess Meredith as the Penguin in *Batman* or Cruella de Vil from *101 Dalmatians*.

The pencil extender almost transforms the writing instrument back into one of its early ancestors – the *porte-crayons* or leadholders used before the wood-cased pencil became so common. Once the lead had become blunt in these instruments, it would need to be removed and sharpened and then replaced. Some pencil holders from the seventeenth century included spring mechanisms to push the lead forward and so could be considered primitive propelling pencils. The first mechanical pencil was patented by London-based civil engineer John Isaac Hawkins and silversmith Sampson Mordan in 1822. Mordan bought out Hawkins's half of the patent and went into business with stationer Gabriel Riddle to produce the 'Ever-Pointed Pencil'. Because they didn't have to be sharpened, the Mordan pencils were much cleaner than their rivals and quickly became popular. Other companies, too, began producing their own versions, although they were mainly marketed as novelties rather than proper writing instruments.

The first person to really establish the mechanical pencil as a serious alternative to the traditional wooden pencil was Charles Keeran from Illinois. His Eversharp pencil, patented in 1915, was a vast improvement on anything else on the market at that time. The magazine inside the pencil could hold a dozen pieces of lead, 'enough for a quarter million words'. Eversharp was bought by the Wahl Company in 1917, and within a few years, 35,000 pencils were being made every day. The history of the Eversharp sometimes becomes conflated with a similarly named mechanical pencil developed at around the same time in Japan by Tokuji Hayakawa. The Hayakawa Ever-Ready Sharp Pencil eventually became known simply as the Sharp Pencil. Although Hayakawa's company later diversified into consumer electronics, its name – Sharp – is a reminder of his early achievement. Japan is still at the forefront of mechanical pencil technology – the Uniball Kura Toga pencil, launched by the Mitsubishi Pencil Company in 2009, features a ratchet system which rotates the lead each time it makes contact with the page, ensuring the point always stays sharp.

Facing such innovation, traditional wooden pencil manufacturers could be forgiven for viewing the mechanical pencil as a threat. However, Staedtler have attempted to imbue their range of mechanical pencils with at least part of their long heritage. The Staedtler Noris range was launched in 1901. With its yellow and black stripes, it remains a familiar sight in classrooms around the world. Extending the yellow and black design to its range of mechanical pencils creates an unusual-looking product – familiar and yet alien. Like the chunky Mini Hatch or the new Routemaster bus, the combination of contemporary design decorated in traditional drag just seems wrong. But it's not only on mechanical pencils that the yellow and black stripes of the Staedtler Noris can be found. Browsing in a branch of GAME one day, I spotted a yellow and black stylus for the Nintendo DS. 'Novelty stylus in the style of school HB pencil,' it explained on the back. There seemed a kind of poetry in this item. The pencil had grown out from the styluses used to write on wax tablets by the ancient Greeks and Romans. And now this tacky £2.99 novelty item seemed to reunite them. Making the connection even stronger, Suck UK offer a 'Sketch Stylus' – a wooden pencil with a hidden secret; 'the "eraser" is actually a built-in stylus for your touch screen devices. Made from ECR (Electro Conductive Rubber) to work perfectly with iPad, iPhone and other devices.'

Pencil, stylus, tablet. The circle is complete.

CHAPTER 5 We all make mistakes

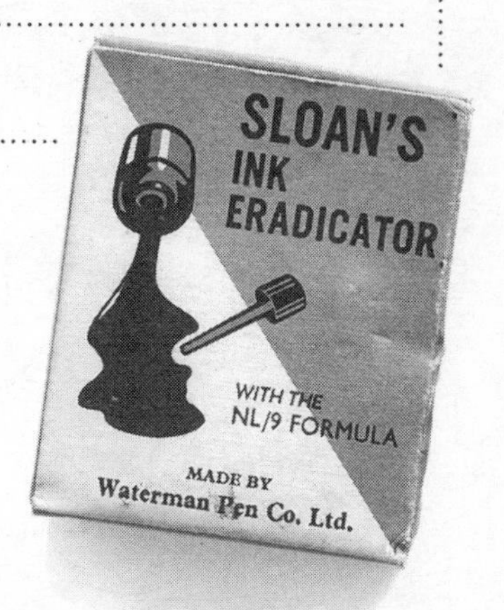

There's a scene in *Eraserhead* by David Lynch where the head of Henry (the main character, played by the tragic Jack Nance) suddenly falls off. The head is found by a small boy who takes it to a factory of some sort. Here, a man uses a core drill to remove a thin cylindrical section of Henry's brain which he feeds into a machine. The machine begins to whirr and a conveyor belt carries a series of pencils along in a row. The core from Henry's head is sliced and fixed to the top of each pencil as an eraser (hence 'Eraserhead'). Eventually, the machine produces a finished pencil and the operator takes one, sharpens it and makes a quick mark on a bit of paper before testing the eraser. The operator nods and says, 'It's OK'; Henry's head makes good erasers and the boy is paid. I've looked into it though, and that's not how erasers are made.

The substance erasers are actually made from has been known under various names (caoutchouc, hevea, olli, kik) for thousands of years. Derived from the 'milk' of various trees and plants in countries with tropical climates, the material was first used by the Olmecs, the earliest civilisation to form in Mexico around three and a half thousand years ago. At this time, the substance was mainly used to produce the large solid balls used in the extremely brutal Mesoamerican ballgame (which would later become known as Ulama). Mixing the latex from the *Castilla elastic* tree with the juice from the *Ipomoea alba* plant (handily the two are often found close to one another), the Olmecs were able to produce thick, flexible strips which could be wound into a ball shape. The material was also used to waterproof fabrics and produce simple artefacts.

In the western world, however, the material would remain completely unknown until the fifteenth and sixteenth centuries, when reports of its unusual properties first started to emerge from the New World. In the middle of the eighteenth century, two French scientists, Charles Marie de la Condamine and François Fresneau, began to see the potential for this 'new' material. La Condamine presented his and Fresneau's studies to the Paris Academy of Science in 1751 – the first academic paper on the subject (published as *Mémoire sur une résine élastique, nouvellement découverte à Cayenne par M. Fresneau; et sur l'usage des divers sues laiteux d'arbres de la Guiane ou France équinoctiale* or 'Memoir on an Elastic Resin Newly Discovered in Cayenne by Mr Fresneau, and on the Usage of Different Milky Tree Saps in Guyane or Equatorial France' in 1755). It wasn't until the later part of that century that the pencil-erasing qualities of this 'gum elastic' began to be exploited.

It seems that British stationer Edward Nairne was the first to realise the substance could be used in this way. In the preface to his *Familiar Introduction to the Theory and Practice of Perspective* in 1770, Joseph Priestley wrote that he had 'seen a substance excellently adapted to the purpose of wiping from paper the marks of a black-lead-pencil' and adds in a footnote:

> It must, therefore, be of singular use to those who practice drawing. It is sold by Mr. Nairne, Mathematical Instrument-Maker, opposite the Royal Exchange. He sells a cubical piece, of about half an inch, for three shillings; and he says it will last several years.

Clearly happy with his purchase from Nairne and pleased with how good it was at rubbing out pencil lines, Priestley is credited with giving the material the name by which it is known today – 'rubber'.

Before that (and indeed for some time afterwards), the preferred method for removing pencil lines was to use stale bread. Even as late as 1846, Henry O'Neill would tell readers of his *Guide to Pictorial Art – How to Use Black Lead Pencils, Chalk and Watercolours* that:

> when a drawing is to be shaded in pencil, the sketch or outline had better be done with a rather soft pencil, in light lines, removing errors with indian rubber or crumb of bread.

However, as the nineteenth century progressed, rubber gradually replaced stale bread as the preferred method of erasing pencil lines, to the relief of all those artists and draughtsmen besieged by hungry ducks.

During the 1830s and early 1840s, the American inventor Charles Goodyear worked on developing a process to stabilise rubber (natural rubber becomes hard and brittle when exposed to the cold; soft and gluey when exposed to heat). By adding sulphur and steaming the material under pressure, he was able to produce something much more durable. However, it was the British Thomas Hancock who in 1844 was first issued a patent for what would become known as the 'vulcanisation' process (meaning 'to put into flames', and taking its name from Vulcan, the Roman god of fire). Goodyear had sent samples of his rubber to British companies to demonstrate its potential before applying for a patent and it seems Hancock basically reverse-engineered an early sample of Goodyear's rubber (having noticed some yellow discolouration which he recognised was caused by sulphur) and filed the patent before Goodyear got around to it. While Goodyear may not have seen any real financial reward for creating the vulcanisation process – he died heavily in debt – his name lives on in the tyre company named in his honour. 'The writer is not disposed to repine, and say that he has planted, and others have gathered the fruits,' Goodyear wrote. 'The advantages of a career in life should not be estimated exclusively by the standards of dollars and cents, as is too often done. Man has just cause for regret when he sows and no one reaps.'

Recognised as a much more durable and reliable material, vulcanised rubber soon began to assert its place as part of the stationery canon, and in 1858, pencil and eraser became one. On 30 March 1858, Hymen L. Lipman of Philadelphia, Pennsylvania was issued US Patent 19,783 for his 'combination of lead-pencil and eraser'. Lipman's design consisted of a lead pencil made 'in the usual manner', except the pencil lead only continued for three-quarters of the length of the pencil – the remaining quarter was instead filled with a piece of rubber:

> The pencil is then finished in the usual manner, so that on cutting one end thereof you have the lead, and on cutting at the other end you expose a small piece of india-rubber, ready for use, and particularly valuable for removing or erasing lines, figures &c and not subject to be soiled or mislaid on the table or desk.

In 1862, Hymen Lipman sold his patent to Joseph Reckendorfer for around $100,000 (around $2.3m today). Reckendorfer later issued his own improvement on the design for which he had paid so handsomely. However, when Eberhard Faber began selling a similar product and Reckendorfer took him to court in 1875, both his and Lipman's patents were dismissed as being invalid. Lipman had simply taken two pre-existing things (a lead pencil and piece of rubber) and stuck them together without producing 'a different force or effect or result in the combined forces or processes from that given by their separate parts'. The court compared Lipman's design to taking a hammer and fixing a screwdriver to the handle, or attaching a hoe to the handle of a rake. The combination of the two might be more convenient, but it doesn't qualify as an invention in its own right that could be patented. In his original patent, Lipman didn't even claim to have invented the idea of 'a lead pencil with a piece of rubber attached at one end' and his repeated use of the phrase 'in the usual manner' in his patent application probably didn't help matters much either.

However, what Lipman's original patent does illustrate is how the world is split into two camps: those who prefer pencils with erasers on the end of them and those who prefer their pencils and erasers to be to separate items. Lipman explained in his patent why the pencil–eraser combination was preferable – it is not 'subject to be soiled or mislaid on the table or desk'. Lipman was arguing for convenience. You make a mistake, and you know the safety net of the eraser is there – just flip your pencil round and it's gone. Personally, I have never liked using the pencil-end eraser. The rubber seems harder and more unforgiving than a normal free-range eraser, and the thought of it cracking or wearing away so that the metal ferrule makes scratchy contact with the page makes me uncomfortable.

There seems to be a US/Europe divide when it comes to the eraser-on-the-end-of-the-pencil debate. In the US, eraser-ended pencils are the default; in Europe they are the exception. But of course it's not as clear-cut as that: Eberhard Faber managed to turn any clean US/Europe division into a smudgy blur. Not only did German-born Faber destroy the claim of America's Joseph Reckendorfer to have invented the 'combination of lead-pencil and eraser', he also launched a standalone eraser which would become a classroom icon in the United States – the Pink Pearl.

The Pink Pearl was designed as part of Eberhard Faber's range of 'Pearl' pencils. A simple pink rhomboid, its distinctive colouring and soft texture were a result of the volcanic pumice mixed with the rubber and fatice during the manufacturing process. Erasers are made from either natural or synthetic rubber, but the rubber itself is just used as a binding agent and typically only makes up around 10 to 20 per cent of the eraser as a whole. Other ingredients are added, including a mixture of vegetable oil and sulphur known as 'fatice'. It is this fatice which acts as the real erasing material. Abrasive substances such as pumice or glass powder are also often added, depending on the texture of the eraser required.

The eraser was launched in 1916, just as compulsory education laws were being introduced across America (in 1918, Mississippi became the final state to introduce such laws). Its low price and reliable quality meant it became a common feature in classrooms across the country and, while it may not be well known in the UK, it is familiar to millions of Americans. In 1967, the eraser was celebrated by the artist Vija Celmins, who produced a series of painstakingly crafted Pink Pearl sculptures from balsa wood, shaped and painted to look just like the real thing. The sculptures took the humble eraser and turned it into the icon it deserves to be – blown up to $6\frac{5}{8} \times 20 \times 3\frac{1}{8}$ and sitting in an art gallery. Ten years later, Avon paid tribute to the Pink Pearl in its own unique way, producing a Pink Pearl nail brush ('Ten busy fingers after school, play and homework need a scrub-away brush to erase undernail dirt!')

The familiar bevel shape and colour of the Pink Pearl are still recognisable today in the version sold by Papermate (the Pink Pearl name has been passed around a bit, appearing originally as the 'Eberhard Faber Pink Pearl', then later as the 'Sanford Pink Pearl' and finally the 'Papermate Pink Pearl'). While the various company names under which it has appeared have changed after each acquisition and merger, the Pink Pearl image itself has remained constant – the 'Eraser' icon in Photoshop, both in shape and colour, is clearly modelled on a Pink Pearl-type eraser. On Etsy today, crafters sell Pink Pearl magnets, Pink Pearl badges and modified Pink Pearl erasers with USB memory sticks embedded in them.

With the development of synthetic rubbers, polymers and plastics during the early twentieth century, erasers could be produced in new shapes, colours and scents. The rounded corners of the Pink Pearl wedge emerged because a sharper corner might break off during transit and it is more comfortable in the hand. A slightly more resilient material can allow for a squarer shape, like that of the clean, white Staedtler Mars Plastic ('Virtually residue-free and with only minimal crumbling') or the Rotring B20 ('It works by rolling up graphite and dirt particles in eraser dust').

The tan Artgum block; the beige Magic Rub; the turquoise Rub-A-Way; a lump of blueish-grey kneadable putty; a white plastic cuboid: erasers come in many different forms. So far, so serious – however, the material's versatility also allows for a little fun. Cheap, colourful and with the pretence of practicality, novelty erasers are the ideal classroom collectable. Shaped like people, animals, household objects (my sister had one shaped like a toothbrush – the handle yellow and the bristles in white), different types of fruit (each given the appropriate perfume; the strawberries smell like strawberries, the snozzberries smell like snozzberries), and the meta stationery-disguised-as-stationery post-modern Ouroboros of the pencil-shaped eraser.

However, despite that versatility, the long rhomboidal form has survived. Sometimes the eraser will be split in two; one smooth pink or white half for erasing pencil lines, and one more abrasive grey or blue half to erase ink. As pencil lead simply sits on top of the paper, removing it is fairly easy. Ink, which soaks into the paper's fibres, is harder to erase. For a long time, the only way to remove the ink marks would be to scrape away at the surface of the page itself. Depending on the type of ink and the type of paper, this could be done with a variety of materials: the abrasive half of a pencil and ink eraser; a piece of pumice stone to remove ink from parchment; or even a metal blade. This was the method I used when working on drawings at university – carefully scratching away the ink from a piece of tracing paper or detail paper with the edge of a razor blade. Once or twice, I'd accidentally nick my fingertip with the tip of the blade and the drawing I'd been working on would be ruined. But if the worst that happened to me was the inconvenience of ruining a drawing with a drop of blood, I was lucky. Early ink erasers in the late nineteenth and early twentieth century looked more like surgical scalpels rather than anything you'd expect to find in an office, and caused much more than a cut finger.

'STABBED TO DEATH IN OFFICE FROLIC' reads the headline from a story in the *New York Times* from 1909. The story explains how fifteen-year-old George S. Millitt of 425 Pleasant Avenue had mentioned to his colleagues that it was his birthday. The girls he worked with began to tease him, telling him that as it was his birthday, he deserved a kiss. 'Every one of them vowed that as soon as office hours were over she would kiss him once for every year he had lived.' He laughed them off, saying that the girls wouldn't get near him:

> As 4.30 o'clock came, and the day's work was over, the girls made a rush for him. They tried to hem him in, and he tried to break their line. Suddenly he reeled and fell, crying as he did so, 'I'm stabbed!'

It seems that as he attempted to evade his colleagues, he accidentally stabbed himself with the blade of his ink eraser. John R. Hegeman Jr, assistant treasurer of the Metropolitan Life Insurance Company where Millitt worked, told the police he was 'quite sure that Millitt's death was a regrettable accident'. It had been Hegeman who had given Millitt the job and it was his understanding that 'he was doing well in the office, and was popular'. Hegeman said that the ink eraser found in the boy's pocket was 'of the regular pattern supplied to employees of the company'. There's a lesson here for all of us: if it's your birthday at work, don't tell any of your colleagues. Stay quiet if you want to stay alive.

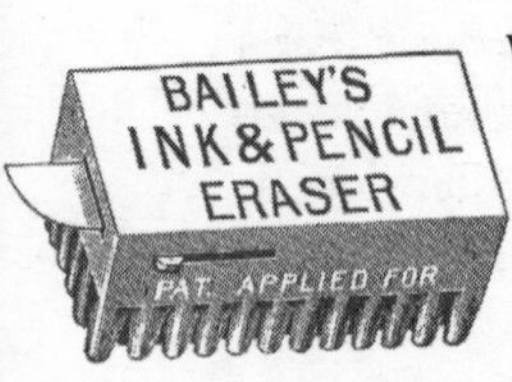

As the typewriter became increasingly popular in the early

twentieth century, so there grew a need to correct typing errors. Harder, more abrasive erasers would be used to remove the typewriter ribbon ink. For accuracy, the erasers were shaped like large flat coins; easier to hold and allowing the typist to pick out one individual letter at a time. Any dust or remnants from the rubber falling into the typewriter mechanism could cause a jam, and so often the disc-shaped erasers would have a long brush attached – a design celebrated by Claes Oldenburg's *Typewriter Eraser, Scale* × currently held at the National Gallery of Art in Washington, DC.

Of course, if you can't completely erase your mistakes, you can disguise them. Hide them. Cover them up. That was what Bette Nesmith Graham did. After leaving school at seventeen, Bette McMurray applied for a job as a secretary at a law firm in Texas, even though she couldn't type. Luckily, she got the job – and the company even paid for her to attend a secretarial school. In 1942, she married Warren Nesmith and the couple had a son, Michael, the following year. The marriage did not last and a few years later, the Nesmiths divorced. Bette was forced to raise Michael on her own, but through a combination of hard work and determination, by 1951, she had reached the position of executive secretary at the Texas Bank & Trust in Dallas. But still she couldn't type properly. For a long time, this hadn't been a problem; she'd just rub out the mistake and retype it correctly. But as the company switched to IBM electric typewriters, she found that the ink from the carbon film ribbons used in these new machines would leave smear marks on the page if mistakes were erased with standard typewriter erasers.

However, Bette found a solution when she volunteered to work overtime one Christmas to help pay the bills. While she was decorating the window for the bank she watched the artists painting the sign boards. 'With lettering, an artist never corrects by erasing but always paints over the error,' Nesmith would later write, 'so I decided to use what artists use. I put some tempura waterbase paint in a bottle and took my watercolour brush to the office. And I used that to correct my typing mistakes.' After a while, her colleagues began to ask Bette if they could have some too, and she realised that there might be some commercial potential in this new product, which she called 'Mistake Out'. Getting advice from one of Michael's chemistry teachers and a local paint manufacturer, Bette made some improvements and paid a chemist $200 to develop a solvent-based formula which would dry quicker than her water-based tempura paint. After changing the name to 'Liquid Paper', and patenting the new formula, she began to market the product beyond her circle of friends and colleagues.

She set up a miniature production line from her garage, with Michael helping to fill hundreds of bottles each month using squeezy ketchup bottles. In 1957, a magazine article about the product boosted sales to over a thousand bottles a month. Despite the increased demand for the product, Bette continued to work at the bank; that is until one day when she was suddenly fired. Ironically, the reason for her sudden dismissal was all down to a typing error. Typing out a letter for her boss to sign, she accidentally typed 'The Liquid Paper Company' instead of the bank's name. The game was up.

Sacked from the bank, Bette now concentrated on Liquid Paper full time, but success wasn't easy – especially as Michael was no longer around to help (in 1965, Michael had replied to an advert looking for 'Folk & Roll [*sic*] Musicians-Singers for acting roles in new TV series' and was now one quarter of The Monkees). In 1968, the company was producing 10,000 bottles a day and grossed over $1m in sales that year alone. Over the next few years, the company grew even more – producing twenty-five million bottles a year by 1975. Four years later, the company was bought by the Gillette Corporation for $47.5m – Bette also received a royalty for every bottle sold until the year 2000. Eventually, however, Bette's fame was eclipsed by that of her son; her innovation and business skills reduced to little more than the answer to a trivia question on a pub quiz machine. After Bette died in 1980, Michael inherited $25m – money which helped him realise his *PopClips* concept of a TV show playing music videos and which inadvertently paved the way for MTV. If video killed the radio star, then it was funded by correction fluid.

While Liquid Paper and Wite-Out are well known in the US, most Europeans would be more familiar with Tipp-Ex. Like Liquid Paper, Tipp-Ex was developed to help correct typing errors, but the original Tipp-Ex product wasn't a correction fluid – in fact the company, led by Wolfgang Dabisch, had been in business for six years before Tipp-Ex launched its first correction fluid in 1965. Originally Tipp-Ex was a corrective tape for use by typists. As described in one of Dabisch's many patent applications, the original Tipp-Ex product was 'a material for obliterating erroneously typed characters from typed paper'. The product consisted of a:

> ... relatively dense base sheet and a covering layer composition. The covering layer is microporous and not substantially penetrated into the base sheet. The covering layer composition is weakly adherent to the base sheet, detachable therefrom by pressure of a typewriter key, and compressible and transferable thereby in substantially the thickness of the covering layer and with substantially sharp contours of the typewriter key.

Essentially, what that means is that it was a bit of paper with some white stuff on it and if you made a typo, then you would go back a space, stick the sheet on top of the paper, retype the letter and the typewriter key would print some of the white stuff over the offending letter and then you'd remove the Tipp-Ex sheet and go back again and then type the correct letter. It sounds impossibly complicated in this age of MS Word, Pages and Scrivener, but this is how things were done back then.

Seeing the success of products like Liquid Paper, Dabisch developed a similar product of his own. Building on the distribution channels he had already created with his Tipp-Ex correction sheets, he was able to establish the Tipp-Ex correction fluid brand throughout Europe before Bette Nesmith Graham even had a chance to move production out of her garage. Dabisch's Tipp-Ex was a huge success, to the point where the product not only became the generic noun for correction fluid in the UK, but it also became a verb. We 'Tipp-Ex' things out, just as we 'Hoover' the carpet – feel free to Google more examples of this sort of thing.

Browsing in Fowlers one day (the same stationery shop where I bought the Velos 1377 Revolving Desk Tidy), I noticed something odd. Behind rows of familiar-looking bottles of Tipp-Ex and other correction fluids (Snopake and QConnect) I saw something I hadn't seen before. I reached to the back of the shelf and picked up two dusty bottles. They were both Tipp-Ex, but they were nothing like the fresh-looking bottles at the front of the shelf. One bottle was a sort of beige – discolouration due to age, or had it always been like that? I looked at it more closely. 'Tipp-Ex Air Mail Fluid' said the label, 'For airmail and light weight paper (art.no 4600)'. The border of the label featured red, white and blue lozenges – like a proper airmail envelope. The other bottle was black. The label for this one said 'Tipp-Ex Foto Copy Fluid. For use on coated or plain paper copies. Will not dissolve toner (art.no 4400)'. The labels on both bottles said, in small writing, 'A West German product', suggesting the two bottles had been on the shelf in Fowlers long before the reunification of Germany twenty-five years earlier. I bought them both, even though the contents had long ago dried up, making them unusable.

Correction fluid was originally (and is still) sold in small bottles, with an applicator brush attached to the lid – similar to nail varnish. But is this the ideal solution? The bottle design has obvious flaws – it can be knocked over, meaning the fluid might spill across your desk. Dried correction fluid can start to collect around the neck of the bottle; the fluid itself can begin to coagulate, the heavy pigment slowly becoming solid over time with the thin, watery solvent sitting on top. As the correction fluid solidifies around the brush, it forces the bristles apart in different directions, destroying any hope of a clean, precise correction. It's messy.

Pentel certainly weren't happy. The Japanese stationery company thought they could improve upon the bottle design. They gathered together a collection of used correction fluid bottles and studied them: lots of the bottles still contained correction fluid, but it had dried and the brush had split; some of them showed signs of spillage. Something had to be done.

In 1983, Pentel launched their new correction fluid bottle design – a small, square bottle with a spring-operated nib built into the lid. Rather than using a brush, the bottle was designed to be turned upside down and the correction fluid applied in a way similar to eyedrop fluid. The tip and bottle shape were refined, and in 1994, they launched a metal-point nib with a pen-shaped body. Who would go back to the bottle now?

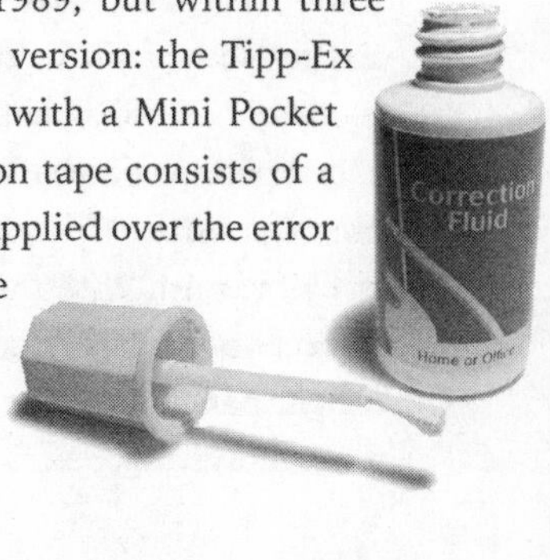

The bottle faced another threat with the advent of correction tape. Correction tape was developed in Japan by eraser manufacturer Seed in 1989. Seed were formed in 1915, and, like Pentel, they were dissatisfied with the conventional correction fluid bottle. However, Seed's new innovation was a dry-tape; if anything making it closer to Dabisch's original Tipp-Ex product. It took Seed around three to four years to develop their product before launching it in 1989, but within three years Tipp-Ex had produced their own version: the Tipp-Ex Pocket Mouse was launched in 1992, with a Mini Pocket Mouse three years after that. Correction tape consists of a white pressure-sensitive tape which is applied over the error using a small plastic applicator (in the case of the Tipp-Ex Pocket Mouse, the applicator is mouse-shaped although that offers no practical benefit). Tape has the obvious advantage over fluid in that it is already dry, and so can be written over immediately. It's also quite difficult to spill.

The bottle had to adjust, but there was limited scope for development. The best Tipp-Ex seem to have come up with is the move from the standard brush to the wedge-shaped foam applicator that 'provides neat and precise correction'.

Rather than just painting or taping over an error, there are more scientific ways of erasing your mistakes. In the 1930s, German pen company Pelikan developed a type of 'ink bleach', originally referred to as 'Radierwasser' ('erasing water') or the more sinister-sounding 'Tintentod' ('ink death'). The name was changed to the more exciting though less explicable 'Tintentiger' ('ink tiger') in 1972, and then was changed again to 'Tinten-blitz' ('ink lightning') two years later.

Two-part formulas such as Sloan's Ink Eradicator (made by the Waterman Pen Company) offered you the opportunity to perform a miniature scientific experiment on the page in front of you. The box contained two bottles (labelled simply '1' and '2'). Using the applicator built into the lid, the user would apply the first solution and then 'agitate until the ink has softened' and use blotting paper to remove the excess fluid. The second solution would then be applied, but the user was instructed to 'not blot until No.1 solution has been applied again'. The product could also be used to remove 'ink, coffee and fruit stains' from white fabric by following a similar process and then rinsing in cold water ('Particular attention is called to the fact that coloured articles should not be treated with this ink eradicator').

In 1977, Pelikan launched the 'Pelikan-Super-Pirat' double-ended pen. At one end was the ink eradicator and at the other was a permanent ink pen with which to rewrite the erased letter or word. With this ink eraser, you only get one chance to correct your mistake, as the eradicator does not work against the permanent ink. The ink eradicator offers no second chances; it is unforgiving.

BASF, 'The Chemical Company', explain how the ink eraser works on their podcast (yes, there's a BASF podcast):

> Let us first take a look at why blue ink looks blue. It contains flat, disc-like colour molecules wherein many free moving electrons buzz around. Light cast upon these electrons is absorbed or 'swallowed' for the most part. Only the blue part of the physical light is reflected. That's why we recognise it as 'blue'.

The ink eraser disrupts these 'colour molecules':

> Now the time has come for the ink eraser. It contains, to a large degree, sulphites that change the composition of colour molecules. Previously flat, they now take on the shape of a pyramid. In this new shape, the molecules cannot move around freely and are no longer able to distribute themselves within the whole molecule. The effect: they reflect all parts of the visible light again.

So although the words remain on the page, they are no longer visible. It sounds like magic:

> What might sound like magic is actually a sleight of hand in chemistry.

This kind of ink eradicator only works with ink of a specific hue – the royal blue used in the majority of fountain pens around the world. The permanent ink of the ink eraser has a different formula and so is not affected by the ink eradicator formula. Pelikan offer this important information about their correcting ink:

> The ingredients of the ink were chosen bearing all possible hazards and threats in mind. Even accidentally swallowing ink would not harm the average person. We do however warn against consumption as ink is not to be considered as nutrition.

That's worth keeping in mind: ink is not to be considered as nutrition.

The ink eraser really only works for royal blue fountain pen ink, but what about ballpoints? Ballpoint ink doesn't react in the same way as fountain pen ink, and instead of turning invisible, it just smudges a bit which makes a mess and ruins the eradicator tip. Instead, a new solution had to be developed.

During the 1970s, while everybody else was [insert glib description of what people were doing in the 1970s], Papermate were busy developing a new erasable ink. The Replay pen (called the Erasermate in the US) was launched in 1979 after a decade of research. The ink in the Replay is slightly drier than the usual thick, viscous ballpoint ink. The different formula requires the ink to be pressurised in order to allow it to flow smoothly and continuously; this means that it can be used upside down, like the Fisher Space Pen, which is handy if you like lying on your back and are prone (no pun intended) to making spelling mistakes. The eraser tip is attached to the lid of the Erasermate, and acts in much the same way as a pencil eraser ('Writes like a pen, erases like a pencil!'), except it doesn't really work very well and seems to produce quite a bit of debris from the eraser tip.

Where once the boundary between pen and pencil seemed reassuringly clear, Sharpie have been working hard to smudge it. Like the Papermate Replay, the 'Liquid Pencil' (launched in 2010) was designed to write 'as smooth as a pen' while erasing 'like a pencil'. The Liquid Pencil had a 'game-changing liquid graphite' which 'eliminates broken pencil leads forever' and promised to 'redefine the way you write'. The properties of this 'liquid graphite' are quite mysterious – even Sharpie don't fully seem to understand what the stuff is. Originally, they claimed that the Liquid Pencil could be erased like a normal pencil immediately after writing, but then the line would become fixed like a pen. However, they later changed this to explain that 'unlike a Sharpie permanent marker, you will always be able to erase it to some degree'. The fact they felt the need to change the wording suggests that if you tried hard enough, you would still be able to erase marks made with a Liquid Pencil even once the line had become fixed. The only limit was your determination it seems. Some inks, however, are so sensitive that they can be erased accidentally.

The Pilot FriXion pen is erased by heat. The thermo-sensitive 'metamocolor' ink becomes transparent when heated above 65°. The ink contains a special 'microcapsule' pigment, consisting of three components: a 'colouring substance', a 'developer to colour' and an 'adjuster for colour change'. At room temperature, this 'colouring substance' connects with the 'developer to colour' and the ink is visible. The FriXion features a small rubber tip at one end, and when the ink is rubbed out with this tip, it creates friction which heats the page. When heated, the 'adjuster for colour change' is activated and combines with the 'developer to colour' and the ink magically disappears.

Because the ink colour is heat sensitive, the product warns you against leaving documents 'near heaters, in a car on a hot day or even putting pages through repeated photocopying' as doing so can cause the ink to turn invisible. Even leaving the pens in the sun can heat the ink inside them. Should this happen and the ink inadvertently turns invisible, Pilot recommend putting the documents (or even the pens themselves) in the freezer, as the thermo-sensitive ink reappears at -12°. It seems that anything written with the Pilot FriXion hovers between the states of 'visible' and 'invisible', depending on the weather. And therefore, for obvious reason, the pack features a warning:

> **Caution**
> This product is not recommended for signatures, legal documents, examination papers or other documents where writing needs to be of a permanent nature.

When it comes to legal documents (cheques, contracts, marriage certificates) you want to know that what you've just signed can only be changed by a lawyer, not by a hair-dryer. While there are obviously lots of legitimate reasons for wanting to erase something – whether it's to correct a simple spelling mistake or just to tidy up an otherwise cumbersome sentence – there are some people who have a more malevolent interest in the possibilities offered by stationery.

Frank Abagnale was probably one of the most successful con-men of the twentieth century. Using a number of false identities (airline pilot, doctor, lawyer), he cashed millions of dollars in forged cheques in the early 1960s before finally being sentenced to twelve years in prison. In 2002, his autobiography *Catch Me If You Can* was turned into the Stephen Spielberg film of the same name, with Leonardo DiCaprio playing Abagnale. Despite the glamour associated with his life of crime, since leaving prison Abagnale has advised banks and businesses on how to prevent fraud. He now travels all over the world, sharing his expertise.

In his book *The Art of the Steal*, Abagnale detailed some of the stationery used by fraudsters to alter laser-printed cheques:

> They take a piece of Scotch Tape – the grey, cloudy kind that doesn't rip the paper when you peel it off – and put it over the dollar amount and over the payee name. They use a fingernail to rub it down hard over the cheque, and then lift the tape off. The dollar amount and the name and the address will come off on the tape. The toner attaches to the Scotch tape and gets pulled from the fibre of the paper. If there's any laser toner residue left over, a little high-polymer plastic eraser will take care of that.

In 2006, Abagnale teamed up with stationery retailer Staples to promote their 'Shred Across America' campaign, highlighting the ways in which people can protect their identity. Perhaps unsurprisingly, the best way to do this, according to the campaign, was by buying a new shredder from Staples. He also worked with the pen manufacturer Uni-Ball to develop the

207 Gel Pen ('The only pen in the world that cannot be altered by chemicals or solvents'). The 207 uses 'specially formulated inks that contain colour pigments. The colour pigments in the ink are absorbed into the paper fibres.' This means the ink is 'trapped' in the paper and cannot be altered, making cheques and documents safer and more secure.

And so finally it seems Abagnale has atoned for his sins against stationery. Although maybe it was stationery which corrupted him in the first place, rather than the other way round. As a teenager, Abagnale worked in the stockroom of his father's stationery store. Could it be that being surrounded by all those erasers and rolls of Scotch Tape gave him the inspiration for his life of crime?

CHAPTER 6

Take me, I'm yours

For some people (myself included) buying new stationery is a joy. Visiting a stationery store, you are surrounded by potential; it's a way of becoming a new person, a better person. Buying this set of index cards and these page markers means I'll finally become the organised person I always wanted to be. Buying this notebook and this pen means I'll finally write that novel. Sometimes, though, people can become a little over-excited when buying new stationery; Morrissey once described visiting a branch of the stationery chain Ryman as 'the most extreme sexual experience one could ever have' (although maybe he meant 'the most extreme sexual experience one could ever have with Morrissey'). Then there are those at the opposite end of the scale – people who never buy stationery, the ones who scavenge for pens and scraps of paper where they can find them. The stationery equivalent of freeganism.

Ever since Argos opened its first stores in the UK in 1973, the retail chain has been closely associated with its famous blue pens. Customers use the pens to write down the catalogue numbers of the products they wish to purchase before queuing up to pay the cashier and then waiting at the designated

Collection Point to receive their items. Everyone is familiar with the Argos pen, but what do we really know of it? I decided to visit my local branch to take a couple of the pens and use them as I would any other pen. It's obvious from looking at it that the Argos pen is designed to be as cheap as possible, and I can imagine that the pen is deliberately intended to be uncomfortable to hold in order to deter people from stealing it. In some ways, this seems rather mean-spirited, but it makes perfect sense. There might even be something to be applauded in such a subtle form of persuasion. Creating a pen so wilfully unattractive that people don't even want to steal it is quite an elegant solution to an otherwise costly problem. It's a sort of practical application of the 'Nudge' theory developed by Richard H. Thaler and Cass R. Sunstein. A modest form of mind control.

As I used the pen, suffering because of its cheapness, I began to wonder how many pens they produce every year. According to their web site, they serve 130 million customers in their stores annually. They must get through a lot of pens, but how many? I emailed them to find out. A few days later, I received a reply thanking me for my enquiry but informing me that the information I'd requested was 'business sensitive' and 'not disclosed to anyone outside of the company'. Would this mean I'd have to get a job at Argos to find the answer to my question? I hoped not, as their head office is in Milton Keynes and that would be an inconvenient commute.

Almost identical in form to the blue Argos pen are the pens used in bookmakers; red in Ladbrokes, lime green in Tote Sport, dark blue in William Hill, dark green in Paddy Power. Each pen looked more or less the same: a straight, thin stick barrel 8.5 cm long, with a 5mm metal nib. If I could find out who supplied the bookmakers with their pens, maybe I could find an answer to my Argos question that didn't involve starting a new job in Milton Keynes.

Tote Sport's pens are supplied by a company called Tate

Consumables. According to their web site, for more than a decade Tate Consumables have been 'a leading UK supplier of Consumables'. I think they're being modest. I'd struggle to name any other UK suppliers of consumables.

> Tate's key areas of operation continue to be within Petrol Forecourts, Betting Shops and all types of Retail Outlets, though the more recent addition of 'Print Services' has extended our operations into many other areas.

However, despite the similarity between the different pens, Tate Consumables do not supply Argos or Ladbrokes, only Tote Sport. 'If they look the same I would say they probably come from the same manufacturer,' explained one of the sales executives at Tate Consumables. I asked him if he could tell me the name of the company who manufacture their pens. 'Sorry I don't have access to this level of information,' he replied. Another dead end.

Looking more closely at the pens from Argos, William Hill, Ladbrokes, Tote Sport and Paddy Power, I noticed subtle differences between them all. Perhaps they weren't the same after all. The Tote Sport pen was slightly rounder, with softer corners; the Ladbrokes pen had edges slightly more defined; the Argos pen more defined still, almost hexagonal in cross-section. I collected a few more samples and began to compare them. There were differences between the pens collected from each chain, but there were also differences between pens from different stores from the same chain. The whole bookmaker thing wasn't helping.

Fourteen years after the first Argos stores opened in the UK, IKEA opened its first store in Warrington. Since then, the Swedish furniture retailer has introduced a rival to the Argos pen in the form of the IKEA pencil. Whereas each catalogue stand in Argos only has slots for six pens and so access to large numbers of them is restricted, the pencils in IKEA are in large dispensers which almost invite you to help yourself to a handful

('We provide the pencils to help customers place orders and we are happy to continue this service,' they explain). In 2004, the *Metro* newspaper claimed that the short wooden pencil stamped with the IKEA logo had become the latest 'must-have' accessory ('actually, make that "must-steal" because millions of the implements given to customers to fill in their order forms are pilfered each year'). The paper reported one customer who was alleged to have picked up eighty-four of the pencils during one trip to the store ('He didn't even buy any furniture' was the inevitable punch-line). The article also noted that 'bingo players on cruise ships in the Med have been spotted marking their cards with them, golfers use them to mark their score cards and some teachers keep a few spare to give out to pupils'.

Just as I wanted to know how many pens Argos produce every year, I had a similar question relating to IKEA pencils. I rooted around their web site and found a document from 2008 celebrating the twenty-first anniversary of the opening of IKEA's first UK store. The document included a page listing twenty-one facts about IKEA. Fact 10 on the list said:

> Last year, IKEA UK customers used 12,317,184 pencils.

I admired the Swedish company's sense of openness. Argos could learn a thing or two from them.

Where Argos appear to have gone for a 'Nudge'-style approach to try to deter people from stealing their pens, banks and post offices have traditionally used a simpler and cruder system: physically attaching their pens to the counter with a metal chain. However, in recent years, even these institutions have been developing their own alternative solutions to the problem of pen theft.

In 2005, Barclays bank piloted a new scheme in five of its branches across the country. As part of an effort to appear more accessible, the black pens on chains were 'outlawed' and were replaced with 'bright blue pens without chains which have messages to encourage customers to feel free not only to use them but to take them home if they wish'. The pens featured messages saying 'Borrowed from my bank', 'Bank swag', 'Take me, I'm yours' and 'I'm free!' The scheme was led by marketing director Jim Hytner, who explained that the chain-on-the-pen symbolised the old relationship banks used to have with their customers: 'Basically we don't trust you to leave this pen behind after you use it, yet we expect you to entrust us with your life savings.' Hytner wanted to move Barclays into the twenty-first century. A press release explained that 'a free pen is a small gesture to show customers we value their custom'. Although, of course, this was way back in the olden days before the banking crisis of 2008 when the offer of a free pen was seemingly enough to make us trust a bank and sign up to an unaffordable mortgage and a lifetime of apparently free and easy debt.

For some, the promise of a free pen alone was too much; they got too excited. While the scheme was still at the pilot stage, four thousand pens were taken in the first five days from a single branch in Bradford. The *Telegraph* quoted a spokesperson as saying 'We expected our customers to take one or two pens, possibly a handful, but some were walking out with an entire box tucked under their arm. We're still going to roll the scheme out nationwide, but we've nailed the box to the counter at our Bradford branch and will probably do that elsewhere.' Undeterred by the Bradford experience, in 2006 the bank rolled the scheme out to all of its 1,500 branches across the country. Ten million pens in the first year, costing the bank 3p a pen. A total of £300,000 worth of pens.

Hytner might be right about the chain on the pen acting as a metaphor illustrating the imbalance of power between bank and customer, but the metal chain presents practical problems as well as poetical ones. This point was highlighted in an email sent to the *Anything Left Handed* web site by a man named David

Dawber who had an unfortunate experience as he tried to send something by Special Delivery at his local post office. 'The pen meant for signing was on the right-hand side of the counter on a short chain. As it would have meant twisting into an awkward position, I used my own pen,' Dawber explained. Dawber then claims that he was challenged by the woman behind the counter who asked why he didn't use the pen provided by the post office. Dawber explained that it was because 'it's on the wrong side of the counter for me, it's discriminating against left-handers'. The situation got worse as the supervisor also got involved, dismissing Dawber's claim and telling him he was 'talking rubbish'. 'I'm not talking rubbish,' Dawber responded. 'It's awkward for a left-hander to write with a pen on the right-hand side, and don't speak to a customer like that.' When he got home, Dawber emailed the Royal Mail to complain about the way he had been spoken to, and to make a suggestion that 'pens are placed in a better position for left-handers'. Dawber is a freedom fighter and should thank Jim Hytner for solving this issue the next time he goes to the bank. But while Hytner believed that the pens in Barclays had nothing to lose but their chains and explicitly invited people to take them home with them, there are other times when the invitation is less clear.

'Hotels invariably place a supply of writing paper in the room. This is meant for the business or social correspondence of the guest,' explains Lillian Eichler Watson in her 1921 *Book of Etiquette*. 'Never take any of the hotel stationery away with you,' she adds. 'It is as wrong in principle as carrying away one of the Turkish towels. Use only as much as you need for your correspondence, and leave the rest behind you.'

But is Watson right? Her view that taking hotel stationery is morally wrong is shared by one concerned reader of the American syndicated *Ask Ann Landers* advice column in 1986. The reader explained that a friend (referred to as 'Mrs Q') travels a lot, often staying in expensive hotels. 'Mrs Q is a great letter writer and I have received many letters from her, always on stationery she took from hotels. I'd be ashamed to let it be known I steal like that. Why would anyone be so stupid?' The letter was signed 'Explanation Wanted in Laredo, Texas'. Landers replied:

> Dear Lar: Hotel stationery is provided with the hope that guests will use it AND take it along. It's good advertising. It's when guests help themselves to towels, bath mats, shower curtains, pictures, pillows, bedspreads, coffee pots and TV sets that the hotels get testy.

A survey for one travel web site found that 6 per cent of the 928 customers they polled admitted to taking stationery from hotels (compared with just 2 per cent who admitted to pocketing miniature jars of jam from the breakfast bar). Is this stealing? Matthew Pack, the CEO of HolidayExtras.com who carried out the survey doesn't think so, saying 'Hotels do factor in the cost of most of these items, and we are actually doing them a favour when we take branded goodies home with us.' The opportunity to spread the branding message on a piece of hotel stationery is often limited – if the hotel is lucky, it will be picked up by someone like 'Mrs Q' who will use it to write a letter to a friend, but much of it must end up tucked away in a drawer somewhere, never to be seen again or just used to scribble down a shopping list or a message while on the phone.

Sometimes though, hotel stationery will reach a much wider audience. Beginning in 1987, and continuing until his death a decade later, the German artist Martin Kippenberger produced a series of drawings made on hotel stationery which he had collected as he travelled around the world. Known as the 'hotel drawings', the works have no real defining theme or style other than the origin of the raw materials. 'Kippenberger grounded his volatile oeuvre paradoxically in a concept of fleeting provenance,' wrote Rod Mengham in an article for the Saatchi Gallery's online magazine. 'His suitcase aesthetic derived clearly from a kind of magnetic aversion to the common understanding of "home"; he reversed the usual polarity and consequently never looked like he was doing anything more than just passing through.' At a recent auction, one hotel drawing ('a self-portrait of the artist, seen with hands clasped behind his back, standing in the corner like a scolded schoolboy') drawn in coloured pencil on a sheet of paper headed with the Hotel Washington logo sold for £217,250.

But if it is true that hotels are happy for us to take their pens as it promotes their brand, how effective is stationery as an advertising or promotional tool? Well, it seems that it is indeed actually pretty effective. Research by the Promotional Products Association International (PPAI) suggests that promotional products perform better than other marketing tools in a number of ways, including 'high recall where the name of the advertiser is remembered', 'repeated exposure to the advertising message because of the length of time the item is kept' and 'a more favourable impression of the advertiser, resulting in a propensity to do business with the organisation giving the item'. The British Promotional Merchandise Association (BPMA) also published similar findings, with 56 per cent of people in one survey saying they felt more favourable towards a brand or company when they received a promotional item. BPMA board director Stephen Barker said that the research showed that 'promotional merchandise is a highly cost-effective form of promotion which gives a ROI [return on investment] that is higher or equal to all other forms of media'.

Obviously, it's in the interests of both the PPAI and the BPMA to highlight the benefits of promotional products, so we shouldn't be too surprised by these findings. But while these organisations represent the interests of promotional product manufacturers in general, there is no reason why they should favour any one particular type of promotional item or product over any other. Yet the BPMA's research quite clearly indicates that stationery is the most effective type of promotional product. In their survey, 40 per cent of respondents had received promotional pens or pencils in the previous twelve months, with 70 per cent of them keeping the products. Other office items ('calculators, staplers, paper pads, rulers, sharpeners etc') had an even higher retention rate; of the 13 per cent who received these items, 77 per cent kept them. Stationery is useful, and people keep useful things – 70 per cent of those who kept the promotional items did so because they knew they would use them. Each time the item gets used, it acts as a reminder of the brand. If the figures are correct, then it would seem that Matthew Pack was right when he claimed we are helping hotels when we take their stationery. But there are also some situations where helping yourself to free stationery crosses the line into good old-fashioned theft.

As you slip a packet of Post-it Notes or a Black n' Red notebook from the stationery cupboard into your bag before leaving work in the evening, it would be hard to claim that it isn't stealing, even if it feels OK somehow. One survey found that two-thirds of people admit to have stolen stationery from work, and while some might acknowledge that what they're doing is wrong, 27 per cent don't even feel guilty about it. When I studied architecture at university, one lecturer (who shall remain nameless) used to reminisce fondly about how well stocked the stationery cupboards were when he used to work for Norman Foster. He helped himself to so many Rotring pens and mechanical pencils that when he finally set up his own private practice a decade after leaving Foster Associates, he was still using equipment he'd stolen from the famous architect's offices.

But theft is theft, and if you get caught, you have to face the consequences. Lisa Smith from Gloucester found this out the hard way in 2010, when she was sentenced to 200 hours of community service for stealing stationery from the Delancey Hospital in Cheltenham. Smith took fifteen boxes of chalk, eight printer cartridges, a packet of batteries, lollypop sticks and toys from the hospital (some of which she later tried to sell on eBay). Although in the eyes of the law what she did was no different to helping yourself to a bottle of Tipp-Ex or a pack of envelopes from the office, it somehow seems worse. Don't take children's toys or printer cartridges, and definitely don't nick them from an NHS hospital, Lisa, that's wrong. You crossed the line.

I still didn't have an answer to my Argos question and my numerous job applications had proved unsuccessful (not a single interview despite applying for over thirty jobs). I had almost given up, but then one day I noticed that the Argos customer service team had set up a new Twitter account, @ArgosHelpers. I tweeted them, asking how many pens are used in their stores each year. 'The answer is a LOT,' they replied, adding 'we are slowly changing over to electronic pads' (and it seems that prior to this changeover, they are also experimenting with pencils instead of pens in some stores). That was too vague for me, so I asked for something a bit more specific. 'Just counted and we ordered 13 m pens last year.' Pretty much the same figure as the number supplied by IKEA. Closure at last.

Chapter 7
Wish you were here

It can't all be work, work, work.

While so closely associated with the office and the classroom, stationery deserves a holiday now and again just like the rest of us. And so it is that when we visit the seaside or a foreign country, so often the gift shops are filled with novelty stationery. Whether it's giant pencils covered in national flags with tassels on the end (you'll often see a pack of mini colouring pencils attached to these tassels creating a hallucinogenic jolt of scale for the bewildered holidaymaker) or pencil sharpeners shaped like local landmarks, retailers will take any opportunity to wrap an item of stationery in a thin cloak of cultural heritage and charge a premium for it. Although, of course, there can be little doubt that wherever you go in the world, the novelty pens you see in the airport gift shops will all have been made in the same factories in some distant country with a questionable attitude to workers' rights.

It's easy to see the appeal. People on holiday want to buy souvenirs, something to remind them of the fun they've had. But at the same time, there can be a reluctance to buy something useless: a cheap plastic statuette or decorative plate.

You don't want to buy something just for the sake of buying something. But a pen is perfect. A pen can be used at home or in the office – and each time it gets used, it's a reminder of the wonderful holiday you had.

In fact, before the holiday is even over, the pen can be put to good use as you write a postcard to send to friends and family back home. As a child, I'd often send myself a postcard when I went on family holidays. It was an act of delayed masochism – I knew I wouldn't receive the card until I got back home and my holiday was over. The postcard was like a time capsule, sent from myself to myself, but the version of myself who sent the postcard was irritatingly smug. 'I'm sitting by the pool,' I'd write. 'I might go for another swim once I finish writing this. Anything good on TV in England? How's the weather?' Back at home, reading this message, I'd reconcile the sense of jealousy I felt towards the version of myself who was still on holiday with the fact that I knew things he didn't. I knew, for instance, that he'd leave his sunglasses behind in his hotel room and that his flight would be delayed on the way home.

But I'm not the only person who sends themselves postcards. In fact, the practice goes all the way back to the invention of the picture postcard itself. The first postcard ever sent, addressed to Theodore Hook Esq, of Fulham, London in 1840, was sent by Theodore Hook Esq of Fulham, London. The hand-drawn picture on the card caricatures the postal service and it is thought that Hook sent it to himself as a joke (there was no internet in those days, people had to make their own fun). Hook's card is also the only postcard known to exist which was sent using the extremely rare Penny Black stamp. The postcard was sold at auction in 2002 for a record-breaking £31,750 to postcard collector Eugene Gomberg. I doubt any of the postcards I sent to myself as a child would sell for quite as much. Maybe fifteen grand, something like that.

Hook's card had featured a hand-drawn image, but as the sending of postcards became more popular towards the end of the nineteenth century, cards began to be printed with images on them: the era of the picture postcard was born. Early postcards would feature etchings and line drawings. These were gradually replaced by prints of paintings and tinted photographs (before the advent of colour photography, black-and-white photographs would be coloured by hand before being printed on to postcards – one would-be artist who hoped to make a living this way was a young Adolf Hitler. Lack of success caused him to pursue other interests). Tinted photographs would eventually be replaced with colour photos following the end of the Second World War.

As rail travel improved and fares decreased more people could afford to make daytrips to the seaside and the picture postcard was the ideal souvenir: on one side, a photograph of the destination or landmark you had visited; on the other, a short message to your friends or family. A picture of the stunning sights you've seen, and a description of the wonderful time you're having. I think postcards are probably more fun to send than to receive.

One way to make the postcard more fun for the recipient is to choose one with a saucy joke on the front. During the first half of the twentieth century, the undoubted 'king of the saucy postcard' was Donald McGill. In his 1941 essay about the artist, George Orwell described McGill's postcards as:

> the 'comics' of the cheap stationers' windows, the penny or twopenny coloured postcards with their endless succession of fat women in tight bathing-dresses and their crude drawing and unbearable colours, chiefly hedge-sparrow's-egg tint and Post Office red.

These postcards would become inextricably linked with the British seaside for decades.

The son of a stationer, McGill was born in 1875 in a middle-class part of London. At school, McGill had been a keen sportsman, but an accident during a rugby match when he was seventeen resulted in the amputation of his left foot. Fortunately, McGill had another talent away from the rugby pitch: he was a naturally skilled artist. He signed up to a correspondence class to learn the art of cartooning from John Hassall (who would later create the famous Jolly Fisherman 'Skegness is SO bracing' poster for the Great Northern Railway Company). Leaving school, McGill spent three years working as a draughtsman for a firm of naval architects before joining the Thames Ironworks, Shipbuilding and Engineering Company.

In his spare time, McGill was a keen painter, and a small exhibition of his more serious paintings caught the eye of businessman Joseph Ascher who licensed the rights to some of his paintings and sold them as cards. Unfortunately, they weren't popular and Ascher was forced to sell the stock as remainders. But when Ascher started selling prints of McGill's 'lighter work' (the saucy cartoons for which he is now remembered), they were an instant success. McGill took inspiration for his cartoons from music hall (his father-in-law owned the Palace of Varieties in Edmonton) and soon he was producing six cartoons a week for Ascher. From this work, he was able to give up his job at the Thames Ironworks and work as a freelance postcard artist. However, even as his designs became more and more popular, McGill still only earned a flat rate for each cartoon. One design (rather tame by McGill's standards) shows a young girl praying by the side of the bed, a dog tugging at her nightdress and the caption 'Please Lord, excuse me a minute while I kick Fido'. The design went on to sell millions of copies, but McGill received just six shillings for the design.

In his essay on McGill, Orwell describes him as a 'clever draughtsman with a real caricaturist's touch in the drawing of faces' but argues that the real value of his work is in the fact that it is so 'completely typical' of the genre he represents. Orwell talks about the 'overpowering vulgarity', 'ever-present obscenity', the 'hideousness of the colours' and the 'utter lowness of mental atmosphere' which is characteristic of McGill's postcards, a proportion of which, 'perhaps ten per cent, are far more obscene than anything else that is now printed in England', leading 'a barely legal existence in cheap stationers' windows'. While for McGill's many fans, the risqué nature of the postcards was key to their appeal; the threat of prosecution for obscenity was ever present during his career.

A series of actions by censorship committees from different local authorities culminated in a trial at Lincoln Crown Court in 1954. McGill was charged with breaking the 1857 Obscene Publications Act for a series of postcards he had published (including one featuring a woman approaching a bookie at a racecourse and announcing 'I want to back the favourite, please. My sweetheart gave me a pound to do it both ways!'). He was given a £50 fine (plus £25 costs) and a court ruling which meant that postcard manufacturers were suddenly a lot more aware of the risks they faced, and several smaller companies toned down their designs (only to be rewarded with bankruptcy as the tamer designs did not sell in nearly the same quantities as their more risqué alternatives, illustrating that the card manufacturers were responding to the desire for low-level smut rather than corrupting an innocent public). In 1957, McGill gave evidence to the House Select Committee as the Obscene Publications Act was being amended. Although the new act was more liberal, ultimately leading to the 1960 publication of *Lady Chatterley's Lover*, the era of the saucy postcard was over. 'I'm not proud of myself,' McGill would admit towards the end of his life, 'I always wanted to do something better. I'm really a serious minded man underneath.'

But McGill's dominance of the British picture postcard scene, with his gaudy illustrations, did not mean that the earlier photographic tradition died away. Indeed, one Englishman in Ireland would help to define the way they would look for many years. John Hinde was born in Somerset in 1916 and raised in a Quaker household. A childhood illness had left him partially disabled, and this, combined with his religious upbringing, meant that rather than signing up to fight in the war, he instead began to explore his interest in photography with the civil defence forces. During the 1940s, Hinde would become one of the key champions of colour photography in the UK. His photographs appeared in a series of books exploring British life – *Britain in Pictures, Citizens in War and After, Exmoor Village*, and *British Circus Life*.

The last of these almost caused him to abandon professional photography completely. After working on *British Circus Life*, Hinde became a circus promoter and met a trapeze artist known as Jutta, who would later become his wife. Hinde even started up his own travelling circus show, and during this time would take photographs of the countryside he passed through. In 1956, he started a postcard company and his brightly coloured photographs of the Irish countryside, filled with stereotypical images of 'whitewashed thatched cottages, red haired colleens and happy, turf-bearing donkeys', quickly became popular. He carefully ensured that every landscape he photographed appeared perfect (and even kept a saw in his car boot so he could cut down rhododendron bushes to stick in front of anything he considered unsightly before taking a photograph).

Returning to England, he began producing postcards for Butlins, and it is for these images of holiday camps that he is best known. These 'highly saturated, colourful images', collected together by Martin Parr in his book *Our True Intent Is All for Your Delight*, represent what Parr describes as 'some of the strongest images of Britain in the 1960s and 1970s'. 'Everything seems exaggerated but oddly everyday,' wrote Sean O'Hagan in the *Observer*, comparing the images to the films of David Lynch or the work of photographer Peter Lindberg. The hyper-real colours and settings of an impossibly idyllic holiday resort, with its 'expanse of blue water, the overhanging plastic plants and fake seagulls, the children bobbing in inflatable tubes overseen by a lifeguard so cool he was wearing shades indoors', still look extraordinary today.

In the same way that Hinde would make sure that every Irish landscape he photographed was perfect (and if reality didn't match his vision, he would rearrange it until he was happy), he would also ensure that every single element in his Butlins photos was correct – even the holidaymakers. In Parr's book, Hinde's assistant Edmund Nägale explains the delicate negotiations involved in making sure each photo was perfect:

> I learnt the basics of diplomacy too – we all know the effect a wide-angle lens has on the human body and the overweight lady in the front row had to go. I would say 'Madam, your photogenic side will feature better in this light if I might ask you to move a bit over here ... bit more ... a little bit more ... and just a tiny little bit more ... Thank you!

For Parr, Hinde's images of Butlins offer 'everything a good photo should. They are entertaining, acutely observed, and have great social historical value.' Parr also observes that 'the most remarkable thing of all is that the cards were painstakingly produced not for any aspirational ideas or as great art, but as humble postcards to sell for a few pence to holidaymakers'. Commercially motivated and ephemeral in nature, the postcard does not look to the future. It aims only to please you here and now – its true intent is all for your delight. Even today, the postcard remains a staple of the gift shop, despite people being able to tweet their holiday photos or update Facebook from the beach. Maybe the cards are now being bought simply as personal souvenirs – not to be sent to anyone but rather to be used as bookmarks, to be stuck on pin boards or refrigerators. A recording rather than a broadcast.

Just as the saucy picture postcards of Donald McGill were not specific to any particular seaside town, so today there survives a bizarre strain of generalised postcard naughtiness. Often these cards will feature a series of men or women in various states of undress. Three naked men on a beach, photographed from behind, with a suggestive slogan saying something like 'Sun, sea and ...' or two topless women sunbathing and the slogan 'Wish you were here!' The function of these cards remains obscure. They seem to attempt the wit of a McGill-type card, but are a sadly literal interpretation of his legacy, making explicit what was once only hinted at.

The most confusing of these postcard images also seems to be one of the most enduring; a close-up photograph of a woman's right-hand breast, except some make-up or other form of decoration has been applied to the breast and as such it has been transformed into a sort of mouse creature. The slogan says 'All the breast from London', which not only fails to acknowledge the mouse-faced breast (is it even a mouse? Who can tell? There is also another version which looks like a fox) but also seems to imply that this is a typical London scene. It isn't. Who came up with the idea? How did they suggest it? Is it meant to be sexually arousing? Surely not. Surely no one could get turned on by that? The card is sold by a company called Kardorama based in Potters Bar, Hertfordshire. Unfortunately, when I contacted the company to find out more about the design, their spokesman was not able to help. ('I am afraid the card you are enquiring about pre-dates my arrival at the business, so I am unable to answer any of your questions.')

It's not just postcards that are sex-obsessed though. The 'tip 'n' strip' pen shows that even something as pure and innocent as a ballpoint pen can be corrupted by man's baser instincts. A pen featuring a picture of a scantily clad woman (or occasionally a man) who, when turned the other way up, becomes even more scantily clad. In one episode of *The Simpsons*, Apu gives Homer one of the pens to keep him occupied during the 45 seconds while his 'Strawberrito' is heated in the microwave at the Kwik-E-Mart. Homer turns to Apu and says, 'You know who would love this? Men.' The 'tip 'n' strip' is just one specific example of the 'Floating Action Pen' (or 'floaty' pen) developed by the Danish manufacturer Eskesen in the early 1950s. Peder Eskesen started his company from his basement in 1946, but it wasn't until a few years later that he developed the product for which his company would become best known.

Floating action pens generally feature a background image (often a landscape view of a riverside or street scene) with a moving object (a boat, car or plane most usually) running in front of it. The window through which we are able to view this scene is filled with mineral oil. Several manufacturers had tried to develop similar pens, but it was Eskesen who found a way of sealing the pens to prevent leakage. Eskesen's first commission was from Esso in the 1950s, who wanted a pen to promote their Esso Extra Motor Oil. An oil drum gliding through clear mineral oil. Perfect synergy (apologies for using the word 'synergy'). Having proven themselves with this pen, other corporate clients began placing orders. Souvenir companies also saw potential in the product and soon the pen became a familiar sight at tourist attractions around the world, with 90 per cent of the world's floaty pens at one point being manufactured by the Danish company.

Of all the items in the stationery cupboard, the Floating Action Pen offers the most realistic way of re-creating your holiday at home. Whether it's a boat gliding along the Seine, a dolphin swimming through the sea or an aeroplane flying over a mountain range; it's almost like being there (if you have a really good imagination). During a trip to New York, I bought a pen that shows a small plastic gorilla slowly climbing up the side of a skyscraper, re-creating with astonishing historical accuracy that time when King Kong climbed up the Empire State Building.

The illustrations in early floaty pens were painted by Eskesen's team of artists. Viewing the image in the oil-filled chamber through the window in the side of the pen causes a distortion; in order to correct this, each element had to be elongated as it was painted, like a floaty form of the entasis principle used by Ancient Greek architects to fool the eye into thinking that the slightly convex columns used in their buildings were actually straight. The illustrations were then photographed, reduced in size and printed on to celluloid film, a process referred to as 'Photoramic' by Eskesen. The small photographic film elements were then assembled inside the pen before the mineral oil was added and the pen was sealed. The Photoramic process was eventually replaced by digital photography in 2006 – a move which caused some concern among Eskesen's more devoted collectors who feared the image quality would suffer.

Unlike the traditional souvenir floaty, the 'tip 'n' strip' begins as a photograph rather than a painted image. A 'mask' is created which, depending on its position, can conceal or reveal specific parts of the image. Like the postcard of the mouse-faced breast, the same photographs of the same models were used for many years. It was only in the mid-1990s that the original 1970s models were replaced. Eskesen still sell the Tip & Strip pen, and have recently introduced a new set of models. The range now includes 'Sarah', 'Rachel', 'Claudia', 'Jennifer', 'Daniel', 'Justin', 'Michael' and 'Nicholas', all photographed with such alarming clarity that even those purists sceptical of Eskesen's switch to digital photography should be satisfied – forget about your office IT department looking at your browser history, these pens are strictly NSFW (not safe for work).

For all their visual entertainment, floaty pens are far from the only type of souvenir pen on offer for globe-trotting stationery enthusiasts. I collected a selection of 'pictorial' and 'sculptural' pens in New York, too, and compared them with the equivalent pens on sale in London.

Pictorial pens represent souvenir pens at their most basic. Essentially, they are nothing more than a pen with a picture printed on the body. Interestingly, the actual physical form of the 'pictorial' pens available in both New York and London is identical (in fact, both were stamped 'KOREA'). All that differs is the imagery printed on the pens themselves. With the pens identical worldwide, it is just the graphics printed on the barrels which gives different cities around the world the opportunity to demonstrate the way in which they view themselves. New York souvenirism is very confident – it's aware of its own iconic status and celebrates itself unapologetically (pens dressed as yellow taxis, dollar bills or the Statue of Liberty). On the other hand, London souvenir pens desperately grab on to any idea of history in an attempt to justify their own existence, regardless of how valid that claim may or may not be. One pen I bought in Bloomsbury celebrates 'Historical London' on the lid, yet alongside images of St Paul's, the Houses of Parliament and Tower Bridge the pen also included pictures of the London Eye and the Gherkin (buildings which can't be much older than the pen itself). One area in which London obviously trumps New York is with regard to royalty. But even in this obviously anachronistic field, the souvenir shops of London still specialise in a desperate kind of nostalgia – the Queen Mother and Princess Diana continue to dominate the scene.

The sculptural pen is defined by the inclusion of a local landmark or figure re-created in moulded plastic perched on top of the pen. Ideally, it helps if the chosen landmark is quite linear in form so as to continue the line of the pen. For this reason, towers and statues are ideal; beaches or lakes are not really suitable. New York, of course, has the perfect sculptural pen icon in the form of the Statue of Liberty. It's almost as if it had been designed to appear on the top of a souvenir pen (it wasn't – the injection moulding process used to produce the pens hadn't been invented in 1886 when the statue was presented to America by the people of France). However, there is one flaw in the design of the Statue of Liberty which impacts on its suitability for this type of pen: the torch. When cheaply produced in plastic, the upraised arm can be fragile. In fact, I bought two Statue of Liberty pens during my trip. The flame of the torch snapped off one. The poor lady's hand snapped off the other.

London doesn't really have anything like the Statue of Liberty which sits as well on the top of a pen. There's Big Ben of course, but that looks a bit odd separated from the Houses of Parliament; Nelson's Column or the Shard aren't iconic enough; the London Eye is too round; Tower Bridge is too wide; and apparently the Angus Steakhouse on Shaftesbury Avenue isn't important enough to justify a pen. Instead, London is forced to rely on its more mundane features for sculptural pens – policeman's helmets and red telephone boxes (the sort which are now only maintained in certain parts of London in order to act as props in the backgrounds of tourists' holiday photos rather than functional pieces of civil infrastructure). It's a sad state of affairs when, as a country, the best thing we have to celebrate in pen form is a phone box and a pointy hat.

But it's not only when we're on holiday that we turn to novelty stationery; whether it's just a desperate attempt to brighten up our desks at work, or when we're running out of ideas for the office Secret Santa, we're grateful to the companies who bravely provide us with these unnecessary products. One such company is Suck UK. Founded by Sam Hurt and Jude Biddulph from the kitchen floor of a north London flat in 1999, Suck products are now available in over thirty countries worldwide. Specialising in gifts and homewares with a humorous twist, the company has a selection of stationery products ('We love stationery and we've made it more fun,' they say, almost suggesting they didn't believe stationery was enough fun in the first place) including 'Mummy Mike' (a silicon rubber band holder shaped like an Egyptian mummy but with elastic bands for bandages); 'Dead Fred' (a pen-holder in the shape of a small man stabbed through the heart with a ballpoint); a tape dispenser shaped like a tape measure; pencils shaped like drumsticks; sticky notes in the style of 3.5" floppy disks and a plastic desk tidy shaped like the lid from a giant BIC Cristal. Things shaped like other things. Never gets tired.

This confusing world of 'things shaped like other things'

becomes even more bewildering when you reach the realm of 'stationery shaped like other stationery'. These range from the novelty pencil-shaped eraser my sister had in her collection to Suck's quite beautiful desk tidy in the shape of a giant pencil sharpener (one of which sits upon my desk as I type these words). The oddest example in my collection is a pencil sharpener shaped like a Pritt Stick. This is a fairly faithful reproduction of a late-1980s Pritt Stick, with the pencil sharpener hidden in the bottom (the lid can be removed to empty it of pencil shavings). Unlike the giant sharpener desk tidy or rubber pencil, there are no clues of scale or texture that could give it away. It really does look like a Pritt Stick. I can't help but feel that the novelty would begin to wear out when, for the fifth time, you accidentally pour a load of pencil shavings over the back of a photo you wanted to stick into a scrap book.

The justification for so much of this novelty stationery is that it is in some way practical. The lie we tell ourselves on holiday as we buy our glittery pens and oversized rubbers is that we can use them when we get home. For the same reason, buying your son or daughter a pen from a tourist gift shop seems better than buying them a toy from the same shop. 'Of course, it's justified because it's a "creative" toy,' complains Jonathan Biggins in his book *700 Habits of Highly Ineffective Parents*. 'Yes, indeed, pencil and paper can be creative tools: if you use them. Simply amassing vast stockpiles of the stuff somewhat defeats the purpose.'

CHAPTER

Back to school

It begins as soon as the schools break up for the summer; you see it in shop window displays, in newspapers and on adverts on TV: 'back to school bargains', 'stock up for September'. A call to action: buy new stationery supplies for the forthcoming academic year. As a child (and despite my love of stationery) I found these adverts upsetting. Looking forward to six weeks off from school with all the possibilities and potential that presented, I didn't want to be reminded that soon it would all be over. But, as the days and weeks of the summer holiday dragged by I'd start getting a bit bored. I'd begin to look forward to life returning to normal. And so soon it was time to stock up for September.

For most high street retailers, December is the key month: that's the month that makes or breaks you. Sales of books, CDs and DVDs bubble along throughout the year, with occasional blips for Mother's Day and Father's Day and other 'gifting events' (a phrase which manages to transform an expression of familial love and generosity into a cynical marketing exercise with utmost efficiency) before spiking in December. Eleven months in the red, and one good month to put you back in the black. But for stationery retailers, Christmas comes three months early, starting with 'back to school' in early September, then graduating to 'back to university' at the end of the month. As soon as the holidays begin, the marketing campaigns are launched; campaigns planned in meetings before the previous year's influx of students had even needed to sharpen their pencils.

When I was at school, the most important thing was the pencil case; everything else came second to that. Your pencil case said a lot about you. In a classroom where uniform restrictions meant you had little opportunity to display your personality through your clothing choices (boys could change the length of their ties, girls the length of their skirts, but that was about it), you had to make the most of the few small gasps of oxygen your individuality could find. Support a particular football team? Get the pencil case. Like a band or cartoon character? Get the pencil case. The licensed products industry is big business and stationery plays a key role in that. However, just how appropriate certain brands are for this kind of licensing is a subject for debate.

I remember having a cylindrical pencil case one year at school designed to look like a can of Pepsi (the white can design used from 1991 until 1998). Another year I had a flatter, rectangular pencil case in the style of a Walkers crisp packet (Ready Salted, although I would have preferred Salt & Vinegar). Looking online now, however, it seems that these kinds of designs are no longer commonly available. I imagine this is due to modern sensitivities surrounding the marketing of sugary and fatty foods to young children, and in retrospect, it does seem odd that schoolchildren (or rather their parents) would pay to advertise fizzy drinks and snacks to other young children. And so perhaps it's good that these products have been discontinued. Well done, society.

However, a quick search online for 'playboy pencil case' depressingly brings up three different designs currently available from one leading online retailer. UK stationery chain WHSmith faced controversy in 2005 when their stores started selling the Playboy stationery range, although they defended the decision saying 'We offer customers choice. We're not here to act as a moral censor.' A statement which could have been viewed as admirable and brave had it not been immediately preceded with the words 'It outsells all the other big brands in stationery by a staggering amount.' WHSmith insisted that the Playboy logo was not 'inappropriate' for schoolchildren, arguing that it's 'just the bunny'. 'It's a bit of fun, popular and fashionable.' However, the Playboy range was quietly dropped by WHSmith a couple of years later. 'We continually review and update our range to offer customers a wide range of products,' a spokesperson explained, 'each spring we renew our range of fashion stationery and as part of this update we have chosen to discontinue the Playboy range.' Where society leads, the pencil case follows.

The majority of fashion stationery is aimed at girls, with a much more limited selection aimed at boys. When I spoke to a buyer at one UK high street stationery retailer, she suggested that this is because girls will often go out shopping together with their friends and if one girl buys a particular item, her friend will buy something different; the girls don't want to have exactly the same things their friends have. A teenage boy, however, tends to feel the opposite – he's less keen to stand out: if his mate has a South Park pencil case, he'll get one too (South Park remains very popular with schoolboys even now, all those years after it was first shown). For boys, there's safety in numbers. The heteronormative assumptions underlying these buying decisions may not make us comfortable, but they appear to be supported by sales figures; in the retail sector, the bottom line counts for more than gender politics. The extent to which the limited range aimed at boys (and gendered market segmentation as a whole) is responsible for shaping the conservative purchasing decisions of the average schoolboy is beyond the scope of this book.

While licensed stationery is hugely popular in the UK, it's not so common in the rest of Europe. Where school uniforms aren't required, pupils are able to express themselves more easily and an Angry Birds or One Direction pencil case loses its appeal. Already, that reference to Angry Birds and One Direction is probably out of date, but it's difficult to future-proof a sentence like that.

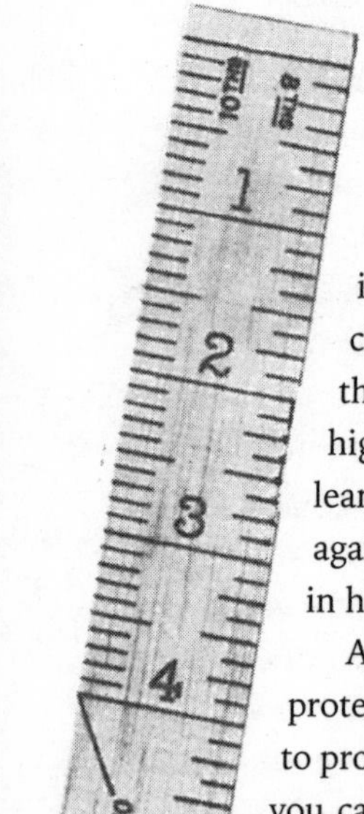

Pencil cases themselves aren't even so common in the US – they certainly don't have the same ubiquity in American high schools that they have in the UK. With lockers more common in the US than in the UK, only the equipment needed for any one particular lesson needs to be taken into the classroom. A quick comparison of the UK and US web sites for Staples reinforces this theory. Even allowing for differences in terminology (not limiting ourselves simply to 'cases', but also including 'pouches', 'boxes' and 'tins'), the UK site clearly has a much wider selection. Suddenly, all those scenes in TV shows and on movies set in US high schools make sense; the love-struck teenage girl, leaning back against her locker, clutching a ring binder against herself, with a pencil and ruler gripped tightly in her hand and no pencil case anywhere to be seen.

And of course, the US gave the world the pocket protector – the plastic pen-holder worn in a shirt pocket to protect from ink stains. Why carry a pencil case when you can keep everything safe in your pocket? Although more commonly associated with people working in stereotypically 'nerdy' fields such as engineering or computing (or people studying such subjects at university) rather than those still at high school, it seems reasonable to suggest that the pocket protector is the natural consequence of a largely pencil-case-free culture; a society looking for some alternative method of stationery transportation. But despite the pocket protector's nerdy associations, there are actually two rivals which each claim the honour of having invented this bit of plastic.

Of the two claimants, Hurley Smith has chronology on his side. His claim pre-dates that of his rival by almost a decade. Smith had graduated from Queens University, Ontario with a BSc in electrical engineering in 1933 but struggled to find relevant work, and spent the next few years trying to market Popsicles to candy stores and grocers. Eventually, he moved to Buffalo, New York and was able to put his education to better use, working for a company producing electrical components. He noticed how his colleagues would often keep pens and pencils in their shirt pocket; this would lead not only to ink and lead stains visible through the white fabric, but would cause the edges of the pockets to fray. At home, he started experimenting with different designs and different plastics, trying to find a solution to the problem. Using a narrow strip of plastic, he folded it in half, and then folded one end down again to form a flap. When inserted into a shirt pocket, the flap would fit over the top, and the longer side would extend slightly beyond it – that way, both the pocket interior and the pocket edge were protected. Smith applied for a patent for his 'Pocket Shield or Protector' in 1943 (the patent was granted in 1947). Gradually, Smith improved his design, heat-sealing the 'pocket shield' along its seams to create a sort of pocket-within-a-pocket. As Smith's design began to sell, he started to target large corporations and businesses looking for unusual advertising opportunities. By 1949, he was earning enough from pocket protectors to be able to quit engineering and focus his attention on his invention full-time.

Smith's rival to the claim is Gerson Strassberg, who demonstrates a less methodical approach to invention. Strassberg (who would later go on to become mayor of Roslyn Harbor in Long Island, New York) was manufacturing clear plastic wallets for bankbooks in 1952 when the telephone rang. As he answered the phone, he put the wallet into his shirt pocket.

At some point during the phone call, he put his pen inside the plastic wallet and the idea grew from there.

In these two rival claims for this one simple product, we see two enduring mythical images of the inventor in action. On the one hand, there is the principled engineer who identifies a problem, then experiments with different materials and designs until finding something which offers an adequate solution. On the other hand, we have the maverick who stumbles across an idea by accident but has the wit and ingenuity to make the most of his serendipitous discovery. It's likely that both these claims for the invention of the pocket protector are legitimate; in reality, half a dozen more people probably invented it too. With plastics becoming more commonplace in the 1940s and 1950s, and pen technology still in its leaky phase, it's not surprising that different people thought of similar solutions to a common problem completely independently of one another.

As attractive as the pocket protector may be, particularly when modelled by someone like Lewis Skolnick from *Revenge of the Nerds*, there is a limit to how much can be carried in a shirt pocket. A few pens and pencils are fine. But if you add a compass, a set square and a protractor, tools used in GCSE maths classrooms every day, then you're going to need a pencil case.

Younger students are luckier. They can travel light. When I was at primary school, I don't think I had a pencil case. We had old-fashioned desks with lids, and would keep everything in there. Not that there was much to keep. Pencils, rubbers and rulers were provided by the teacher. It was only once we had upgraded from pencil to pen that we were expected to provide our own equipment. Moving from pencil to pen was an important step; it was a promotion which could only be earned once the teacher was satisfied with your ability to correctly write in joined-up writing. A divide formed within our class – those still on pencil (invariably the black and yellow Staedtler Noris HB), and those, such as myself (not that I mean to boast or anything) fast-tracked on to pen.

After making the leap from pencil, the first pen many schoolchildren are introduced to is the Berol Handwriting Pen. The red-barrelled pen was designed after a detailed consultation with teachers to determine which characteristics they thought were essential for a child's first pen. The thickness of the barrel and the amount of resistance on the page were considered important for a child getting to grips with a pen for the first time; the pen needed to be relatively chunky for clumsy, inexperienced fingers; it shouldn't glide across the page like a ballpoint would. Launched in 1980, the Berol Handwriting Pen has remained a classroom favourite (although it's hard to say if this is because of the specific qualities of the pen itself, or as a result of nostalgia: for many teachers, it would have been their first pen too).

Now a subsidiary of Newell Rubbermaid, Berol started life as the Eagle Pencil Company. Eagle was founded by the Berolzheimer family in New York in 1856, with its first London office opening in 1894. The company opened a factory in Tottenham in 1907, and would remain in the area for the next eighty years until it relocated to bigger premises in King's Lynn. As the company grew, it acquired other brands (including pencil manufacturer L. & C. Hardtmuth in 1964 and educational art material producer Magros Ltd in 1967). With this increased product range, the old name of Eagle Pencil Company no longer made sense, and the name was changed to Berol Limited in 1969.

As well as the Handwriting Pen, in the school classroom Berol were associated with the felt-tip. Building on the experience of Magros Ltd, Berol were able to dominate the classroom colouring-in market just as they also dominated so many children's first experiences of handwriting pens. But for all its innocent fun, the world of colouring-in can also be deeply political. Which colours are included in a pack – and more importantly, the names given to those colours – reveals a lot about society.

In 2004, Berol introduced a new 'Portrait' range of felt-tips. 'Increasingly, pupils have been frustrated by the lack of colours available to reflect their skin colour and those of the pupils around them,' Berol explained in a press release when the range was launched. The new range was intended to allow pupils 'to represent a wider range of hair and skin colour in the classroom, perfect for portraits and drawings of people'. The Portrait range included six new colours: Mahogany, Peach, Olive, Cinnamon, Almond and Ebony. By naming the colours after neutral objects (woods, fruits, nuts, spices) rather than anything directly related to skin tone, Berol avoided the possibility of causing offence. But some companies have not always been so forward thinking.

The Binney & Smith Company was formed in 1885. Originally created as the Peekskill Chemical Works Company by Joseph Binney in New York in 1864, the company was taken over by Binney's son, Edwin, and nephew Harold Smith on Joseph's retirement. At the end of the nineteenth century, Binney & Smith began producing school supplies such as slate pencils and dustless chalk. In 1903, the company launched its first set of eight crayons: Black, Blue, Brown, Green, Orange, Red, Violet and Yellow. As more colours were added to the 'Crayola' crayon range, Binney & Smith were forced to become more creative with their colour names. The 1949 colour range included 'Burnt Sienna', 'Carnation Pink', 'Cornflower', 'Periwinkle' and, unfortunately, 'Flesh' (used to refer to a pinkish, whiteish tone). Within a few years, the company rightly realised that this name might be slightly problematic and so, as Crayola explain, it was 'voluntarily changed to "Peach" in 1962, partially as a result of the US Civil Rights Movement'.

While the problems with using the name 'Flesh' to represent a peach colour are pretty obvious to us today, there have also been times when the colour naming team at Crayola could perhaps be accused of being over-sensitive. Introduced in 1958, 'Indian Red' was renamed 'Chestnut' in 1999 after concerns that schoolchildren in the US believed that the name referred to the skin colour of Native Americans. In fact, according to Crayola, 'the name originated from a reddish-brown pigment found near India, commonly used in fine artist oil paint'. Although it was an innocent name, I think Crayola were right to make the change. Perhaps learning from their mistakes, when they launched their range of coloured pencils in 1987, they gave them safer names: 'Red', 'Red-Orange', 'Orange', 'Yellow', 'Yellow-Green', 'Green', 'Sky Blue', 'Blue', 'Violet', 'Light Brown', 'Brown' and 'Black'. The only possible source of controversy in that list is their sudden decision to use the slightly ambiguous 'Sky Blue' rather than 'Light Blue' which would be more in line with the other names they've used and less likely to cause confusion (what time of day? What time of year? The sky changes colour constantly).

Although artists had been using coloured chalks and pastels held in *porte-crayons* since the middle of the seventeenth century, the first wooden coloured pencils were developed by Yorkshireman Thomas Beckwith in 1781 who patented a method of making a coloured compound 'prepared from the purest mineral, animal, and vegetable productions' mixed with 'a proportionable quantity of the most pure impalpable mineral calx or the choicest terrestrial fossil substance'. 'Due proportions of well dephlegmated essential oils' and 'animal oleaceous substances' were added, and the mixture was combined until it was of a smooth texture. It was then rendered 'by the means of mollifications, inspissation and exsiccation by fire' before being encased in wood and formed into pencils. Modern manufacturers tend to use a mixture of pigments, water, binders and extenders to make the pencil core, as terrestrial fossil and animal oleaceous substances can be rather hard to come by. Beckwith's 'New Invented Coloured Crayon Pencils' were produced by stationer George Riley in London in 1788 and could be used with all the convenience of 'a Black Lead pencil' and none of the mess and dust associated with chalk. The coloured pencils were available in thirty-two different shades; today the Swiss company *Caran d'Ache* offers 212 shades across their range of products. Colour inflation.

As you progress through school, the range of items which fill your pencil case increases and diversifies. But once you leave, there are certain items of stationery, which unless you go on to have a very particular career (architectural draughtsman or naval officer for instance), you will simply never use again. For years, we carry these items around with us every day, and then all of a sudden, our relationship with them is severed. There should be some sort of ceremony to mark this occasion; some way of saying goodbye. Instead, they are just put into a drawer and forgotten about. Kept, because they are still perfectly functional, but simply never used. Hundreds of thousands of abandoned set squares and protractors sit idle in desk drawers up and down the country; a great many of those set squares and protractors no doubt manufactured by the same company.

The company that would one day become Helix was founded by Birmingham businessman Frank Shaw in 1887. The son of a goldsmith, Shaw was exposed to the metal industry from an early age. His Hall Street Metal Rolling Company initially began producing steel wire and sheeting before moving into the manufacture of laboratory equipment; test tube holders and clamps, as well as brass compasses. With the increasing numbers of children attending school in the late nineteenth century, Shaw quickly became aware of the growing demand for educational equipment. At the time, a lot of school equipment (such as rulers, set squares, drawing boards) was made from wood. In 1887, Shaw opened a factory for his newly established Universal Woodworking Company and the company quickly became known for the quality of its rulers, which were manufactured in a twenty-two-stage process. 'Fifteen-foot half-round logs were delivered to the factory yard where they matured for up to a year,' Patrick Beaver explains in his book celebrating the first hundred years of the company's history. 'They were then reduced by power saws to ruler blanks of various sizes and these were matured for another year.' After this second maturing period, the blanks would be shaped by machine, then polished and lacquered. A die-cut process would mark out the numbers and graduations on the rulers, which were rubbed with carbon-black before finally being cleaned and polished once again. Finally the rulers were ready to meet their public.

Impressive as the Universal Woodworking Company's twenty-two-stage ruler manufacturing process undoubtedly was, they were at least able to benefit from the processes developed during the Industrial Revolution. The people who made the very first rulers were not so lucky. And yet they were still able to produce standardised measurements of incredible accuracy. Of the very earliest known rulers, the example found at Lothal (in what is now part of the state of Gujarat in western India) is the most remarkable for its accuracy. Discovered during an archaeological dig in the 1950s and dating back around four and a half thousand years, the ivory scale that was uncovered is accurate to around 0.005 of an inch. Ivory and bone would continue to be used for the manufacture of rulers and scales well into the nineteenth century as it was strong and would not warp or twist out of shape, but for reasons of cost and practicality, wood and metal eventually became more common. Some rulers would combine the two – my Velos 145 ruler (another eBay purchase) features a boxwood body, with a steel 'spine' running along one edge to ensure the ruler is able to provide a perfectly straight edge without any of the nicks or dents to which wooden rulers are vulnerable.

Buoyed by the success of his boxwood rulers, Frank Shaw merged Hall Street Metal into Universal Woodworking in 1892 and the company focused increasingly on educational supplies. Two years later, he developed a new compass design: the Helix Patent Ring Compass. Until Shaw's invention, a small screw was used to hold the pencil in place when using a compass to draw a circle, but this meant that a screwdriver would be required to remove or adjust it. Instead, Shaw's compasses used a metal ring which could be tightened or loosened by hand. A new age was born: one in which everyone could draw a circle on a piece of paper slightly more easily than they could before.

The metal compass can also be used as a weapon in any outbreaks of classroom warfare – David Bowie's unusual eyes are often said to be the result of a fight involving a compass with school friend George Underwood when they were both fifteen, although this story appears to be apocryphal ('It was just unfortunate. I didn't have a compass or a battery or various things I was meant to have – I didn't even wear a ring,' Underwood explains in the Bowie biography *Starman*).

Beyond the spike of the compass, an even more fearsome classroom weapon has to be the scissor. In order to reduce the danger posed by arming a roomful of children with bladed instruments, scissors with rounded 'blunt tips' are often used and strict instructions against running with scissors given (and of course, you should always pass scissors by the handle). While you might think that the apparently contradictory idea of 'safety scissors' is the product of the over-protective, health- and safety-obsessed culture of the last few decades, in fact the concept goes back a little further than that. Amos W. Coates of Alliance, Ohio filed a patent in 1876 for his 'improvement in scissors' for children ('more particularly for the use of little girls in cutting out their little quilt-patches, doll-babies &c') which featured 'two blades with two terminal bulbs or guards' to prevent the child from 'running the points into themselves or falling on the same'.

In 1912, Shaw combined a number of his products when the company launched its first mathematical set for schoolchildren. The maths set included 'a Helix 5 inch compass, boxwood protractor, wooden set-square, 6 inch ruler, pencil and rubber'. It was an immediate success, not just in the UK but also with missionaries in Africa and India, and was the precursor to the Oxford Set of Mathematical Instruments ('Complete &

Accurate') still available from Helix today. Despite being on the shelves of stationery shops for over a century, the contents have barely changed and over one hundred million sets have been sold to date around the world. The modern set, in its tin decorated with a drawing of the Chapel and Old Library at Balliol College, contains the same items as the earlier version (though the ruler, set square and protractor are now plastic) but adds a second set square (30°/60° in addition to the 45°/90° original), as well as a pencil sharpener, lettering stencil and a little fact sheet listing mathematical formulae and symbols. These items are all illustrated on the back of the packaging, although be warned, 'Due to individual market requirements, contents may vary from the items shown'.

Approaching his sixties, Frank Shaw began to think about retiring. As he had no children to take over the business (he didn't get married until he was in his late fifties), he sold the business to two trusted associates: Arthur Lawson and Alfred Westwood. Lawson was Shaw's solicitor, and Westwood had been with the company for twenty years and managed the factory. Lawson and Westwood each paid £5,000 for their half of the company (the equivalent of around £220,000 each today). When they bought the company in 1919, its turnover was around £17,000 a year; Lawson and Westwood were able to increase this to £25,000 within two years, but growth stalled during the 1920s and 1930s.

In 1925, Arthur Lawson's son Gordon and Alfred Westwood's son Cliff both joined the company. Gordon would become responsible for woodwork production, with Cliff working his way up to foreman of the metalwork factory. Five years after joining the company, both sons were appointed to the board of directors. To cope with the economic problems following the Great Depression, the company rationalised the business, dropping the least profitable product lines and freezing salaries. Overseas expansion during the latter half of the 1930s allowed the company to recover slightly, but the outbreak of the Second

World War didn't exactly help things. Production turned to serving the war effort, focusing on precision measurement and navigation equipment. Following the war, the company began to concentrate on expanding the overseas part of the business, particularly focusing on the Commonwealth countries that were building and developing their education systems. Gordon Lawson became chairman in 1952 when his father died, and two years later, Cliff Westwood sold his shares in the business to Lawson and retired. At the same time, Lawson's wife, Elsie, joined the board. Helix was now a family business.

Throughout the 1950s, the company began to modernise. Readymade celluloid set squares and protractors were bought in from outside companies and wooden production was slowly phased out. The company name was changed from Universal Woodworking to the Helix Universal Company Ltd in 1955. Four years later, the company opened its first plastic moulding facility. The future was here at last. As timber costs increased and the reliability of the plastic moulding manufacturing processes improved, plastic rulers had become increasingly desirable; lighter and cheaper than their wooden counterparts, plastic rulers are also transparent, making them perfect for technical drawing. But there was one drawback: they weren't strong. Early plastics were extremely brittle. The rulers could shatter with just one tiny jolt.

Improvements in manufacturing processes created rulers which were less likely to break. Helix's rulers were now 'Shatter Resistant', two special words familiar to anyone who went to school in the UK at any point in the last thirty years. The words still appear on their rulers today, as if this was a quality unique to Helix products and not the industry standard. In every classroom, there is always at least one pupil who misunderstands these words and thinks they are claiming the ruler is unbreakable (and then attempts to prove the claim incorrect). But the words are more modest, the ruler is simply 'resistant' to shattering. They're not claiming that it's impossible to snap the thing in half. The typeface 'Shatter' (which was designed by Vic Carless in 1973 and won him that year's Letraset International Typeface competition) has jagged lettering, reminiscent of broken glass, and illustrates the difference between something shattering and merely snapping in half. I'd try making this argument to my classmates at school, but they wouldn't listen, choosing instead to test the claim printed on my ruler. Some rulers upgraded the strength of their claim from 'Shatter Resistant' to 'Shatterproof'. The ruler on my desk (stolen from a previous job) combines these two ideas and claims to be 'Shatterproof Resistant'. I'm not sure exactly what this means. Is it resistant to being shatterproof? I am in constant fear of it spontaneously exploding into a million tiny pieces at any given moment. This is no way to live. I'd throw it away, but I'm not sure it's safe to pick it up.

The flexibility of the plastic ruler means that it not only resists shattering, but can also be used to catapult erasers across the classroom, or produce that satisfying 'thwuppp' noise as it vibrates against the edge of a desk. Some rulers, however, are not only physically flexible, but functionally flexible too. None more so than the 'Combination Letter Weigher & Ruler' sold by Parva Products in the 1940s ('The Xmas gift that's different'). The ruler featured a notch at one end, and three holes in the middle; to weigh a letter, you would slide it into the notch and insert a pencil into one of the three holes. If it all balanced, you knew the letter matched the weight marked next to that hole and could calculate the postage accordingly. As well as allowing the user to weigh their letters before posting them, the ruler also acted as a magnifying glass, French curve, compass, protractor, spirit level and set square.

In November 1959, Gordon Lawson died suddenly and Elsie became chair of the company, with their son Peter becoming general manager. For the next ten years, Elsie would travel across Africa, Asia and the Middle East promoting Helix products, as well as joining British trade missions and the

British National Export Council. Her hard work paid off. By the end of the 1960s, the company had contracts in eighty countries and she had an OBE. During the 1970s and 1980s, the company would continue to expand, purchasing cashbox manufacturer Dunn & Taylor in 1975 and, two years later, eraser manufacturer Colonel Rubber Limited (whose name makes it sound like it was founded by a reject from a stationery-themed edition of Cluedo).

Just as Frank Shaw had found, as he faced retirement, that he hadn't left quite enough time to have a successor, so the fifty-year-long Lawson dynasty ended with chairman Mark Lawson, who had taken over from his brother Peter. There had been some speculation about the future of the company as Mark approached his late sixties, and in January 2012, it was announced that Helix had entered administration. Managing director Mark Pell denied reports that the move was the result of financial difficulties faced by the company, instead explaining that the step was needed as the 'antiquated' structure of the old family business was brought up-to-date. Whatever you say, Mark. Whatever you say.

Less than a month after the company entered administration, it was announced that the French stationery manufacturer Maped would buy Helix. Maped were formed in France in 1947, originally as *Manufacture d'Articles de Précision Et de Dessin*. Initially, the company manufactured brass compasses, however they eventually began to diversify. In 1985 the company started selling scissors and in 1992 added erasers to its range with the purchase of French eraser manufacturer Mallat. Throughout the 1990s, the company added staplers, pencil sharpeners and other school and office accessories to its product range. As Maped grew, it began to add international subsidiaries, starting with the formation of Maped Stationery Ltd in China in 1993 and continuing with new operations in Argentina, Canada, the US and the UK. Maped and Helix had both found success with the same product – the brass compass – before expanding into other school and office equipment, and so it made sense for Maped to buy its British competitor. As the company's president Jacques Lacroix said when the announcement was made, 'There is a remarkable synergy between both companies regarding product portfolio and geographic penetration.' Like Jacques Lacroix, I am also 'pleased that Helix will be able to preserve its operational autonomy, while benefiting from the Maped Group's strong manufacturing and commercial support', although to be honest, I can't imagine Frank Shaw or Elsie Lawson talking about 'remarkable synergy' or 'product portfolios'.

CHAPTER 9

The highlight of my life

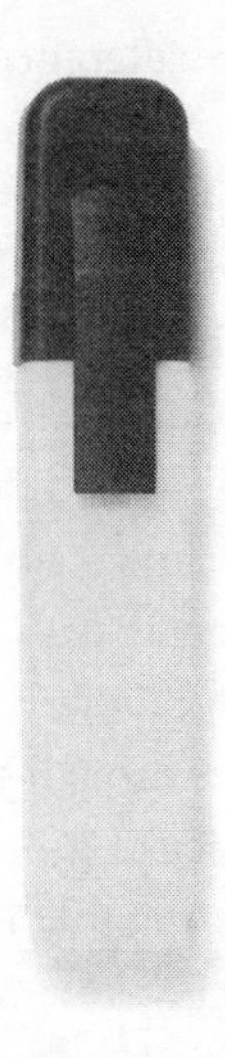

It almost seems unimaginable now: a life before highlighters. Once upon a time – not even all that long ago – if you wanted to emphasise a key word or important detail in a document, you'd have to underline it. A red pen might stand out against black ink, but a thin wobbly line was still the best that could be hoped for. What the world needed was a pen with a chisel tip, equally suited to picking out one small word or a full paragraph, filled with a bright transparent ink which wouldn't obscure or smudge the text beneath. A highlighter. But before the highlighter could be produced, the right tip was needed.

It was Yukio Horie in Japan who first developed the fibre-tip pen that would eventually give birth to the highlighter. The fibre-tip (or 'felt-tip' as it is often called) acts in much the same way as a brush, soaking up the ink and then dispensing it on to the page; the one obvious difference is that with the felt-tip, the reservoir of ink is held in the barrel of the pen rather than in a pot or artist's palette. In 1946, Horie formed Dai Nippon Bungu Co. (which would go on to become the Japan Stationery Company, and later, global giant Pentel), originally manufacturing crayons and writing brushes for use in education. Seeing the growing success of the ballpoint, Horie decided to develop a new type of pen himself, hoping to create something unique to allow his company to stand out. Horie wanted something that wrote like the brushes used to produce Japanese script, but with the convenience of a ballpoint.

By taking a bundle of acrylic fibres and binding them together with resin, Horie was able to produce a tip hard enough to be shaped into a fine point, yet soft enough to easily soak up the ink. Fine channels within the tip allowed the ink to flow to the end through capillary action. The ink itself had to be thin enough to flow through the channels, yet viscous enough to prevent it from leaking. A tiny air hole in the barrel allowed air to escape, ensuring that the air pressure inside did not increase when the pen became warm, causing a leak. After eight years of development, Horie's new pen (called a 'Sign Pen' because the strong line and ease of use made it ideal for signing important documents) was ready. The pen was initially met with indifference on its release, but after Horie took it to the US, it grew in popularity. At some point, one managed to find its way into the White House and was used by President Johnson. It was named *Time* magazine's Product of the Year for 1963 and was even used on board the Gemini space missions.

While it's true that other companies had produced felt-tip pens of some sort or another for years – Lee W. Newman filed a patent for a marking pen with an absorbent tip ('In practice I prefer to fashion it from felt') in 1908, and Sidney Rosenthal invented the Magic Marker in 1952 – Horie's innovation of using fibres bound together with resin meant the tips could be much finer and more accurately shaped, so paving the way for the chisel and bullet tips we use today. Horie's contribution was noted by the *New York Times* which, in 1965, wrote about the growing popularity of these fibre-tip pens. 'There is little argument about the role of a Japanese manufacturer in bringing the market for such a writing instrument to life,' the newspaper wrote. 'The Japanese company, with headquarters in Tokyo, is known as the Japan Stationery Company. Its marker is known as the Pentel Pen, by now a widely used item in the United States homes, schools and business offices.'

The *New York Times* article explains how American firms quickly saw the potential of these pens and started developing their own, noting that 'virtually every major United States pen or pencil manufacturer has moved into this rapidly growing field':

> Parker has joined such other blue-chip competitors as the W.A. Sheaffer Pen Company; Scripto Inc; Esterbrook Pen Company; Venus Pen and Pencil Corporation, and Lindy Pen Company. To these and other familiar pen and pencil makers there must also be added the perhaps less familiar names of such companies as Speedry Chemical Products Inc (maker of the Magic Marker), Carter Ink and others, all of which are active in the broad-beamed marker field.

There's something incredibly uplifting about reading a long list of separate independent companies like that, rather than the small number of conglomerates and their subsidiaries that would comprise such a list today. From that list, every company has either closed down or been absorbed into another company (Esterbrook and Venus merged in 1967 to form Venus Esterbrook which was in turn bought by a subsidiary of Newell Rubbermaid, which also owns Parker. Sheaffer is part of BIC and Crayola now own Speedry's Magic Marker brand).

As finely shaped fibre-tip pens spread across America, new types of inks and pigments were also being developed; lighter water-based inks didn't soak into the page as much as alcohol-based inks. At the same time, developments in pigmentation meant that bright colours like yellows and pinks could be produced – bold enough to leap from the page, yet at the same time transparent, allowing the text underneath to remain visible. The highlighter's time had come.

While Carter's Ink may not have been familiar to the *New*

York Times, the company was well established as a manufacturer of inks. William Carter started the company in Boston in 1858, renting a commercial property from his uncle. Originally known simply as The William Carter Company, it sold paper to local businesses. As he expanded the business, however, Carter started buying inks in bulk and re-bottling them under his own name. This rather ingenious business plan was brought to an abrupt halt by the outbreak of the American Civil War. Carter had been buying ink from a company called Tuttle & Moore, but when Tuttle joined the army, Moore wound down the business. Carter then began to manufacture his own inks, having licensed the formulas from Tuttle & Moore on a royalty basis.

Carter took on new premises to give him space for manufacturing equipment and was joined by his brother Edward, changing the company's name to William Carter & Bro. The name wasn't future-proof though. Shortly afterwards, they were joined by their other brother, John, and so the name was changed to William Carter & Bros. In 1897, their cousin joined the company and the name was changed once again, this time to Carter Bros & Company. If William had just gone with something generic in the first place, he could have saved a fortune in letterhead design. Carter Bros & Co. changed its name one final time to Carter's Ink Company and continued to develop new products throughout the early twentieth century – carbon paper for typewriters, pens, typewriter ribbons, new types of inks. Always innovating to maintain its success.

In 1963, seeing Pentel's success with the fibre-tip Sign Pen, and with their expertise in ink development, Carter launched a new product: the Hi-Liter. Originally only available in yellow, the pen was priced at 39c (around $2.99 today). One advert in *Life* magazine explained:

> **CARTER'S READING HI-LITER** – Clear, 'read-through' brilliant yellow lights up words, sentences, paragraphs, telephone numbers, anything. 'Hi-lite' them so you can find them again fast! Dries instantly, doesn't penetrate paper.

The pen was advertised alongside Carter's other new markers: Carter's Marks-A-Lot ('Addressing packages? Identifying tools, toys, boots, boxes? Marks-A-Lot does it best – boldly, clearly, permanently'), and the Glow-Color Marker ('Like to pep up posters and decorations? Put some "sizzle" into signs and displays? Get this exciting new marker in 5 different, flaming, fluorescent colors that dazzle the eyes – make striking unusual effects!') The advert also included the all-important call to action: 'Look for all these Carter's markers at your favorite store. Buy several today!'

The Carter Hi-Liter was a success and continues to sell well in the US. The pen is now available in a wide range of colours, although yellows and pinks continue to dominate the highlighter market, representing around 85 per cent of total sales. Sitting right in the middle of the spectrum of visible light, yellow leaps out from the page and can be seen more easily than any other colour (even by people with red-green colour blindness). The Hi-Liter introduced a new way of note taking, a new way of revising, a new way of studying. It's probably an exaggeration to claim it changed the world, but only a slight one. Yet just as the world needs revolution, so it needs evolution. There were some people who weren't happy with the Hi-Liter. One such person was Günter Schwanhäusser.

Schwanhäusser was visiting the US in the early 1970s when he saw one of these new highlighter pens during a trip to a stationery shop. He made it his habit to visit stationers whenever he left his native Germany to ensure he was always up-to-date with the latest trends in pens and pencils around the world. But this wasn't just a curious obsession or idle fascination; Schwanhäusser had stationery in his blood.

In 1865, Günter's great-grandfather, Gustav Adam Schwanhäusser, acquired the Grossberger & Kurz Pencil Factory. Opened just ten years earlier by Georg Conrad Grossberger and Hermann Christian Kurz, the factory in Nuremberg had quickly fallen into debt and so was a risky investment for twenty-five-year-old Gustav. However, within a few years, he managed to turn around the fortunes of the ailing company. Ten years after acquiring the factory, Gustav patented a process to manufacture copying pencils. A copying pencil was a pencil which also contained an aniline dye; when a letter written with the pencil was moistened and pressed against another page, a duplicate copy could be made, albeit one in mirror image. Very thin translucent paper would be used so that the printed copy could then be reversed and read normally.

The Schwanhäusser factory grew over the next few years, as did the size of their pencils – at the 1906 Bavarian State Exhibition the company unveiled what was then the largest pencil in the world (at around 30 metres in length, it would seem pretty puny compared to the current World Record holder – the 225-metre-long monster created by Staedtler in 2011. 'The pencil was sharpened at the top and was used to write some words on a piece of paper in front of the notary,' states the Guinness World Records web site solemnly).

In 1925, the Schwanhäusser factory (by now shortened to 'Schwan') introduced the brand, if not the product, for which it would become best known – STABILO (which, right from the beginning, was always written in upper case for extra impact). The original STABILO product was a colouring pencil with a 'remarkably slim but exceedingly firm lead that releases colour in a delicate, velvety soft manner'. This new lead, developed by Dr August Schwanhäusser (the younger son of Gustav Adam and Günter's great-uncle), was both thinner and stronger than other colouring leads used at the time, and so added 'stability' to the product (hence 'STABILO'). The new lead was so much stronger than others on the market that Schwan advertised it with the slogan 'The pencil that never breaks'. This slogan drew the attention of the Association of Pencil Manufacturers, who challenged it. A compromise was reached, with the slogan changed to 'The pencil that doesn't break' (although as far as I can tell, that means the same thing – the Association of Pencil Manufacturers gave up their fight too easily).

By the time Günter Schwanhäusser joined the company in 1950 (becoming the fourth generation to join the family business) the STABILO range had been expanded to include high-quality pens and pencils. Two more ranges had been added alongside it: Othello (aimed at the mass consumer) and SWANO (aimed at children). The company had also diversified into cosmetics – initially launching an eyebrow pencil in 1927 and later adding lip-liners and eyeliner pencils (despite not being particularly well known in the UK, the Schwan Cosmetics section still accounts for just over half of the group's total sales each year).

Just as American companies had been quick to pick up on the emerging fibre-tip market, so Schwan began to introduce similar pens into their range. In the late 1960s, Schwan introduced two fibre-tips: the STABILO OHPen (for use on acetate overhead transparencies) and the STABILO Pen 68 ('the first fibre "painting" pen for school and hobby'). When his father and two uncles retired in 1969, Günter and his cousin Horst took control of the company. Horst looked after the cosmetics side of the business while Günter concentrated on developing new writing instruments. It was during this time that he saw his first highlighter. Günter thought that the idea had potential, and he could see how popular the pen was with American students revising for exams. He thought he could extend that popularity into the larger and much more lucrative office environment, but he wasn't particularly impressed by the quality of the pens themselves. The yellow ink was 'dirty', and the shape of the pens themselves ('a simple round shaft with a clunky felt tip') didn't excite him either. He felt that this new type of pen shouldn't look like anything else on the shelves at that time. It should be unique. Günter believed that if he found a way to overcome these problems, his new highlighter would be a success. He returned to Germany with a sense of purpose.

The first issue was the ink. Without the right ink, the highlighter would be no better than the examples he found in America. Chemistry had always played an important role at Schwan. Dr August Schwanhäusser (Günter's great-uncle) had developed the thin-core colouring pencil which formed the foundation of the STABILO brand and August's son, Erich, continued the tradition, earning a PhD in chemistry before joining the company in the 1920s. As Günter had no direct background in chemistry himself, he gave the company's Head of Research & Development, Dr Hans-Joachim Hoffman, the task of producing a new, brighter, fluorescent ink.

Fluorescent inks and paints had originally been developed by Californian brothers Robert and Joe Switzer in the 1930s. After an accident unloading crates at work, Robert fell into a coma for several months. After he awoke from the coma, he found that his vision had become impaired and was advised by his doctor to stay in a darkened room until fully recovered. During these dark moments in his life, he became interested in ultraviolet light and the properties of fluorescent and phosphorescent compounds; things that glow in the dark (all his sensitive eyes could cope with). Once he had recovered, the two brothers began to experiment with the fluorescent materials they discovered in the storeroom of their father's drug store (they'd take a UV light into the storeroom and anything that glowed, they used). Mixing the naturally occurring fluorescent compounds they found with shellac and other materials, they eventually produced a fluorescent paint they called 'Day-Glo'. During wartime, Day-Glo paints and dyes were used by the military – troops could be identified from the air using fluorescent fabrics, and the paints allowed the US to continue flying from aircraft carriers throughout the night as runway markings would be visible even after dark, giving them the edge over the Japanese forces. After the war, the paints were used for road markings and traffic cones, for fire exit signs in buildings, and for clothing and leisure products. During the 1960s and 1970s, the use of ultraviolet light (or 'black light') became popular with hippies who used pens such as the Carter's Ink Glow-Color Marker to design vivid psychedelic posters. Hoffman's team developed these kinds of inks for use in Günter's highlighter.

As I read through a copy of the company's history, you can probably imagine the sense of satisfaction – the visceral thrill – I felt as I ran the chisel tip of a yellow STABILO highlighter over the words 'the experienced chemists at the company, under the direction of Dr Hans-Joachim Hoffman, were soon able to produce a luminous fluorescent yellow ink'. What greater tribute could be paid to Dr Hans-Joachim Hoffman and his team than to use that exact same luminous fluorescent yellow ink they developed to highlight their greatest achievement?

But despite developing a luminous fluorescent ink for his new highlighter, Günter still had a big problem to solve. He wanted the pen to look and feel different. He wanted something unique, not just another round barrel with a lid on the end. Something new. He asked his team of designers to come up with a new design. They came up with ideas, but nothing was right. They modelled different designs and concepts out of clay; thin and thick, short and long, searching for inspiration. Eventually they settled on a conical cylinder shape – a round barrel thicker at one end than the other. Convinced they'd found the kind of design Günter was after, they showed him the clay cylinder, but he still wasn't happy. Frustrated by this impasse, one of the designers slammed his fist down on the clay model and squashed it flat. Günter liked it.

Despite being designed by accident, the 'fat and flat' shape of the highlighter is actually ideal for the way it is used. Its unique shape means you can identify it without taking your eyes off the page and the flat wedge shape means it won't roll off the desk. It's solid and chunky and reassuring. The designers took this flattened cone shape to the engineers, who produced a mould. The body of the pen would be produced in the same colour as the ink – so just as the highlighted words would leap out from the page, the pen itself would leap out from the shelf in a crowded stationery store. While the colour of the pen body would match the ink inside, each pen was given an identical black twist-cap to create a uniform image. By cutting one corner off the chisel tip, an extra neat little feature was created – a variable line width; using the thick end of the tip gives a 5 mm line (suitable for highlighting large chunks of text), flipping the pen over gives a 2 mm line (for picking out individual words).

Günter was convinced that his new highlighter would be a success, but there was one final thing missing: the pen needed a name. It needed to be something as iconic as its new shape; something that showed the importance of the pen; something short, something powerful. BOSS. The STABILO BOSS. It sounded good. In fact, Günter thought it sounded better than good. He liked it so much he decided not to give the pen a product number like everything else the company produced: STABILO BOSS was enough on its own.

Now happy with the ink, the shape and the name of his pen, Günter Schwanhäusser was ready to launch it into the world. He was aware that the new pen was a hard sell – people were used to using a normal pen to underline text, why would they want to pay extra to buy a special pen that did something they'd never needed to do before? It wasn't just a new pen he needed to sell, it was a new behaviour. He worked strategically, sending samples of the pens to a thousand influential figures in Germany; business leaders, university professors, company owners, even the German chancellor. Each sample was accompanied by a letter. 'It helps simplify your work, saves precious time for more important things and belongs on your desk,' he wrote. Having established in the minds of business leaders throughout the country that the pen was an invaluable aid to productivity, he started sending it to middle-management and departmental heads. These were the people who would place large orders if only they would share Günter's vision of a fluorescent future. Fortunately, after using the pen, they did. During the development of the STABILO BOSS, Günter Schwanhäusser had left nothing to chance. 'No product launched since then was as thought-through as that one,' Günter would later write. The hard work paid off, and the STABILO BOSS would go on to become the most popular highlighter in the world.

The success of the BOSS highlighter reinvigorated the company, and allowed them to open a new production facility in Weissenburg specialising in plastic injection moulding. The new factory meant the company could experiment with new shapes and forms. Where previously they had devoted their attention to developing new inks or leads, now they were able to concentrate on ergonomics – how the pens felt in the hand – creating pens specifically for left-handed people and shaped rubberised grips for children. In 1976, due in no small part to the success of the BOSS, the company changed its name from the Schwan Pencil Factory to Schwan-STABILO.

While highlighters had been enough to cement the fortunes of the Schwanhäussers, sadly the same could not be said for Carter's Ink, the company that had launched the original highlighter back in 1963. Carter's was bought by Dennison Manufacturing (who would later merge with Avery International to become Avery Dennison). The Hi-Liter is still part of the Avery Dennison range and is still a common feature on desks through the US, but sadly little of Carter's Ink remains – at some point during the Dennison acquisition, the company's records (including early contracts and formulas for inks which were never fully developed) were lost.

In the years following its initial launch in 1963, the highlighter (in its various forms) has continued to face different problems and challenges. The greatest of these challenges has been the frequent changes in office and domestic printing technology. Each new ink, each new printing method, produces new problems to be solved. To a highlighter manufacturer, nothing could be worse than that murky smear caused by the highlighter spreading the ink of a printed document across the page. Sometimes, to avoid that smudgy horror, the manufacturer will be upfront and alert you in advance to their limitations. I have a box of S510F Pentel See-Thru Markers I bought from eBay, which states that the pens are 'not recommended for use on NCR copies' ('NCR' referring to a type of carbonless copy paper developed by NCR Corporation), though what I like most about those pens is the way that the phrase 'See-Thru Marker' suggests an entire alternate reality where the word 'highlighter' never took hold.

While STABILO have continued to develop their pens to keep up with printing technology (the STABILO BOSS EXECUTIVE launched in 2008 features a 'unique patented anti-smudge ink with special pigments' and is 'tested and recommended by leading printer manufacturers such as Lexmark'), there have been some attempts to avoid ink altogether in the pursuit of creating the ultimate smudge-free highlighter: a pen-free highlighter experience. Staedtler, Moleskine, STABILO and others offer 'dry' highlighters – thick-leaded, fluorescent pencils producing a strong, bright line on the page. Alternatively, there is the highlighter tape – a transparent, fluorescent tape which can be written on immediately. The tape, launched by Henkel in 1999, has the added advantage that it can be removed once no longer required. More recently, Pilot have added an erasable highlighter pen to their FriXion range – the FriXion Light.

You don't even need to have the printed document to highlight text in it. The Highlight tool in Microsoft Word is specifically intended to 'make text look like it was marked with a highlighter pen' (and of course, the default colour is yellow). We no longer live in a world where we need highlighters to highlight things. Anything is possible.

CHAPTER

10

I'm sticking with you

‘I’m sticking with you, ’cos I’m made out of glue,’ sang Moe Tucker on ‘I’m Sticking With You’ by the Velvet Underground in 1969. It’s a nice sentiment but one which makes little sense. If Moe were made out of glue, surely she’d stick ‘to’ me rather than ‘with’ me. If anything, the use of the word ‘with’ instead of ‘to’ suggests that I’m the one made out of glue. I am not made out of glue. Around the same time as Moe Tucker was claiming to be made out of glue, a German company were also thinking about the sticky stuff. But they gave it a bit more thought.

In 1967, Dr Wolfgang Dierichs, a researcher working at German manufacturing company Henkel, went on a business trip. He checked in and boarded the plane. He took his seat, fastened his seatbelt and got ready for take-off. By the time the plane landed, Dierichs had an idea that would go on to revolutionise the world (of glue). At some point during the flight, he saw something that inspired him: it was a woman. The woman in question was carefully applying her lipstick, and as Dierichs watched her, he began to think that the lipstick form could have a different application. You could take that design, a thin twistable tube, and fill it with a stick of solid glue. It would be clean and convenient. You'd just remove the lid and apply as much as you needed. No pots, no brushes, just a stick of glue. Of course, most people, seeing a woman apply lipstick wouldn't think 'Imagine if that was glue she was smearing over her lips', but Dierichs worked in Henkel's sizeable adhesives division and so it's (almost) understandable that he made that connection.

While the official Pritt Stick history credits the woman on Dierichs's flight with providing the inspiration for his invention, sadly her identity remains a mystery. In the various versions of this story told on their web site, in press releases or in their company history, the woman is never named. It's not clear whether she was a fellow passenger or a member of the cabin crew; she's just described as 'a woman' – sometimes 'a young woman', sometimes 'a beautiful woman'. A cynic might question whether or not the woman ever existed, or if Dierichs was even on a plane when he had his brilliant idea. Perhaps the whole scenario had been created years later to add a degree of romanticism to the whole thing. But I am not a cynic. I believe Henkel.

Fritz Henkel was twenty-eight years old when he and two colleagues formed Henkel & Cie in Aachen, Germany in September 1876. The company initially started selling a washing powder made from sodium silicate (or 'water glass'). Although similar products were already on the market, Henkel's version was more convenient – rather than being sold loose, Henkel sold theirs pre-packaged in small packets. Two years later, the firm began selling a type of bleaching powder; Henkel's Bleich-Soda would be the first German brand-name detergent. The company began to grow, opening new factories and employing sales staff who would travel across the country to promote their range of products. New and varied lines were added; not just cleaning products, but also hair pomade, beef extract and loose-leaf tea.

Over the course of the next century, the company would continue to operate in various business sectors, eventually focusing on three main areas: 'laundry and home care' (including brands such as Persil, launched in 1907); 'beauty and personal care' (Schwarzkopf hair products); and 'adhesives, sealants and surface treatments'. This last was Dierichs's department. Henkel originally began developing adhesives during the First World War, when a shortage of the materials required to produce the glue used in their packaging production led them to experiment to see whether the same material they used for some of their detergent products could be used as a substitute.

Humans have been sticking things to other things in different ways for hundreds of thousands of years. Stone flakes, stuck together with birch-bark-tar to make primitive tools, found in central Italy in 2001 are believed to date back around 200,000 years. Bark from the birch tree was heated to produce the thick glue-like tar. Bitumen was also used to produce early stone tools. Some examples found in the Sibudu Cave in South Africa, dating back 70,000 years, show traces of compound adhesives; rather than just using naturally occurring substances like bitumen as a glue, red ochre would be mixed with plant gum to 'haft' pieces of stone on to wooden handles to produce weapons. The red ochre strengthened the joint (plant gum on its own is rather brittle and would shatter on impact) and made the gum less likely to dissolve in damp conditions.

In his 1930 book on the history of glue, *The Story of an Ancient Art*, Floyd L. Darrow describes how glue was used to produce veneered furniture in ancient Egypt. A mural from Thebes, dating back three and a half thousand years, shows a glue pot warming on a fire and a workman applying the glue with a brush. Furniture found in tombs from the time shows clear evidence of glued joints and veneered surfaces. In *The Story of Furniture* (1904), Alfred Koeppen and Carl Breuer state that 'The Ancient Egyptians already employed the art of ennobling an ordinary wood by gluing on thin layers of costly woods.' Koeppen and Breuer describe the two types of glue used by the Egyptians:

> the common kind, joiner's glue made from animal offal and fish bladders – and also the glue made from caustic lime and white of egg or casein.

As well as using veneer for decorative purposes, the Egyptians found it also had structural benefits – veneered furniture would warp less than if it had been made from a solid piece of wood. Furniture from the period has survived to this day, showing the skill of the Egyptian craftsmen (it's unlikely that my wobbly IKEA bookcase will survive quite as long). 'Historical research shows that the rich beauty of this furniture would have been wholly lacking had it not been for the mastery of the art of veneering,' writes Darrow, 'and in this ancient form of artistry, glue holds the key position.'

The art of glue-making was well known to the Greeks and Romans too. In *Naturalis Historia,* Pliny the Elder credits Daedalus with inventing glue, and describes two different types: bull-glue and fish-glue. Although collagen (which itself comes from the Greek word *kola,* meaning 'glue') can be extracted from pretty much any bit of any animal to make the gelatine used in glue, 'the best glue is that prepared from the ears and genitals of the bull,' explains Pliny. Fish-glue was known as 'ichthyocolla', taking its name from the fish from which it was made. Pliny also describes a 'common paper paste' made of 'the finest flour of wheat mixed with boiling water and some small drops of vinegar sprinkled in it', similar to the flour and water paste still made in primary school classrooms to this day.

While Pliny mentions Daedalus creating glue from bulls, the animal most closely associated with glue production is the horse. Retired horses are sent to the 'knacker's yard' (a term which is believed to have originated in the sixteenth century in Scandinavia and which comes from the Old Norse word *hnakkur*, meaning 'saddle') to be turned into glue. But there is no reason to suggest that horses make better glue than any other animal – they aren't particularly sticky. In the UK and US, horses are used as working animals rather than being bred for their meat. And so, instead of ending up on our plates, a different use for them was found once their working lives were over.

Following the end of the Roman empire, the art of veneering largely died off until it gradually re-emerged during the sixteenth and seventeenth centuries: the first ever commercial glue factory opened in the Netherlands in 1690, producing glue from animal hides. In 1754, a man named Peter Zomer registered the first patent in Britain for his method of producing glue from 'the tails and fins of whales and from such sediment trash and undissolved pieces of the fish as were usually thrown away as useless and of little or no value by the makers of train oil' ('train oil' was made from whale blubber and took its name from the Middle Low German *trän* and Middle Dutch *traen*, meaning 'tear', as it was extracted in droplets and was used in oil lamps and in the production of soap). The glue, therefore, was an economically and environmentally sound use of the by-products of this manufacturing process. Although as Zomer himself admitted, the resultant fish-glue was 'inferior to the glue made in England from the cuttings of leather'.

Glue manufacture remained fairly localised until the late nineteenth century, not least because it was difficult to develop a product with a shelf-life long enough for it to be transported and sold without drying out. William LePage was one of the first to find a solution to this problem. By using sodium carbonate in the production of his fish-glue, he was able to remove all traces of salt, 'which had been the great obstacle to complete success heretofore'. Previously, the fish skins had been descaled, but LePage left the scales on, securing 'a material advantage in retaining in the glue all the useful properties of the scales, which render the glue more insoluble and stronger than is the case when the skins are first descaled'. LePage's glue remained liquid for months after production and so was ready to use straight from the container – other glues had to be heated before use.

LePage had been born in 1849 in Prince Edward Island, Canada, but moved to Massachusetts to work as a tinsmith and later as a chemist. The area had a large fishing industry; much like Zomer before him, LePage used discarded by-products to make his glue. The LePage Company was formed in 1876 and initially sold its produce to local leather goods manufacturers. In 1880, the company launched a glue suitable for domestic use. Over the next seven years, LePage's company would sell 47 million bottles of their glue worldwide.

In 1905, Frank Gardner Perkins from Indiana developed a vegetarian alternative to the meat and fish options on the glue menu, producing the first successful vegetable adhesive. A few years earlier, Perkins had been involved in a disastrous attempt to introduce cassava (a South American crop) into Florida. The crop was not suited to the climate of Florida, as the investors in his Planters' Manufacturing Company spent $300,000 finding out for themselves. During his trials with cassava Perkins noticed that flour produced from the crop would sometimes develop a glue-like consistency if it got damp. Unable to grow cassava, Perkins began to import the flour and experimented with the glue. Once he was satisfied with his formula, Perkins approached the Singer Manufacturing Company and attempted to convince them of its merits, with the hope that they would start using it in their factories.

Vegetable starch had been used to produce glue for thousands of years, but it had never really been used to manufacture furniture. Singer wanted to test the glue for themselves. They assembled a number of wooden cabinets and 'placed them in every disadvantageous place imaginable – on the tops of boilers, behind steam radiators, in the basement, in rooms subject to marked changes in temperature, and elsewhere'.

The cabinets were left in their 'disadvantageous places' for over a year before being thoroughly examined. Fortunately for Perkins, the glue held firm. With the backing of Singer, Perkins continued to refine the formula for his glue. Starting with Perkins No.1, it took him until batch No.183, two years later, before he was happy.

Although vegetable glue was relatively cheap to produce, and despite the advantages of LePage's liquid fish-glue, glues made from cattle and horses still dominated the market. With resources scarce following the outbreak of the First World War, Henkel began their experiments with 'water glass' (the sodium silicate they used in their cleaning products). Initially, this wasn't intended as a new commercial opportunity, but as a way to solve their own packaging issues. Progress was slow, and it wasn't until 1922 that the company began to produce reliable and commercially viable adhesives. To begin with, Henkel concentrated on decorator's glue, including *Henkel-Kleister-trocken,* a dry paste which was soluble in water. In the 1930s, the company also began experimenting with cellulose adhesives. Production continued throughout the Second World War, with some foreign civilians and prisoners of war replacing the employees who had been called up for military service. Following the war, the company also began to develop synthetic resins.

The first adhesive based on synthetic resin to achieve commercial success had been produced in 1922 by Alex Karlson, a shoemaker from Sweden. Just as Peter Zomer and William LePage had used the by-products of local industries to produce their fish-glues, Karlson also used what was around him to produce his glue. He took leftover celluloid from the Swedish film industry, and melted it with acetone to produce his multi-purpose adhesive. As well as providing the raw materials for Karlson's glue, the Swedish film industry was also used for publicity purposes. Karlson's associate Olow Klärre introduced the company's mascot, a donkey named Peppo, in grand style. Klärre covered Peppo in banners advertising the company and hijacked a local parade that was being filmed by a news crew. Newsreel footage of the parade was shown in cinemas throughout the country – was this the birth of guerrilla marketing?

By the 1960s, Henkel had produced a range of successful synthetic adhesives and resins. They had also begun to establish themselves in the US, acquiring Standard Chemical Products Inc in 1960. Then, in 1969, the company launched the Pritt Stick. Within two years, Pritt Stick was available in thirty-eight countries around the world and today it is sold in over a 120 countries worldwide. Around 130 million Pritt Sticks are produced every year and more than two and a half billion have been sold since the product was launched ('Enough to leave a line of adhesive extending from the earth, past our satellite the moon, on to Mars and then all the way back again' the company claims).

The Pritt Stick is now advertised by a character called 'Mr Pritt'. Mr Pritt is a Pritt Stick which has somehow come to life. He has no origin story, like Peter Parker being bitten by a radioactive spider or Bruce Banner being exposed to gamma radiation; he just appeared in 1987, emerging fully formed from the mind of Arnold Sindle from Manchester-based advertising agency Boden Dyhle Hayes (BDH). The body of Mr Pritt is red, like the international version of the glue stick (the body of the UK version is white). But on the red sticks, the lid is also red. The lid of Mr Pritt is white. Is 'lid' the right word? It's his head. Or perhaps the lid is a sort of helmet, with his head underneath? It's not clear how it works. Regardless of Mr Pritt's bizarre physiology, every time you use a Pritt Stick, you are effectively smearing the sticky brains of one of Mr Pritt's children on to a bit of paper.

Since 2007, the red labels on the sides of Pritt Stick tubes have included an image of Mr Pritt himself (he only began appearing on the white UK design in 2011), although as graphic designer Randy Ludacer of Beach Packaging Design in New York observed on his web site, 'the character that appears on the label appears to be wearing the *old* version label (in which he, himself, is *not* pictured)'. If Mr Pritt were to update the label on his body to more accurately reflect the current label design (a design which his very presence made obsolete), then we would enter into the world of the recursive Droste effect – the Pritt Stick featuring a picture of Mr Pritt who in turn features an image of a Pritt Stick featuring a picture of Mr Pritt and so on, repeated into infinity. 'Of course,' points out Ludacer, 'since so much of Mr Pritt's lower half is cropped out of the label, it's debatable how many iterations are possible.'

But the Pritt Stick isn't the only glue stick on the market. Once Dierichs's idea was turned into reality, other brands were quick to launch their own versions, though none has managed to create anything with the same level of impact as Henkel's Pritt Stick. There have been many attempts to find some new way of marketing a stick of glue in a plastic tube. Glue sticks are sold as stronger than their competitors or as lasting longer than a rival's product. Coloured dye is added to some glues which then magically vanishes as the glue dries ('Goes on purple, dries clear!'). In such a crowded market-place, you have to admire the restraint shown by UHU who promote their Glue Stic with the simple boast that it features a 'unique screw cap to protect from drying out' (even though preventing the glue from drying out is the very least you would expect the lid to do).

UHU was formed by German chemist August Fischer in 1905, after he acquired a small chemical factory in Bühl. In 1932, Fischer developed a synthetic resin adhesive which was completely clear and, whereas previous glues like Karlson's Klister were just 'multi-purpose', Fischer's formula was 'all-purpose'. It could stick anything, including early plastics like Bakelite ('UHU sticks everything better'). The glue was named after the onomatopoeic nickname for the eagle-owl then familiar to the nearby Black Forest ('Don't say glue … say "Yoo-Hoo"').

Still, where the UHU Glue Stic struggles against the mighty Mr Pritt, UHU do have one advantage over their rivals: the UHU Glue Advisor app (available on both iPhone and Android) allows you to select two different materials and then tells you which sort of glue is most suitable to stick them together. Want to stick a natural pearl to some lead? Use *UHU plus Endfest 2-K Epoxidharzkleber* (the English language app links to the German product range, unfortunately). Sticking a cork on some concrete? Use *UHU Montagekleber Universal*. But there are some situations where the UHU Glue Advisor can't help you. Sometimes you want to stick something together, but a glue isn't appropriate. You wouldn't use a Pritt Stick to repair a torn twenty-pound note, or to wrap a birthday present. Tape. You need tape.

Adhesive tapes of various forms have been used for centuries. The Egyptians made funerary masks using strips of linen soaked in plaster. Diachylon, a mixture of various plant juices with olive oil and lead oxide, was applied to linen bandages for use in medicine by the ancient Greeks. In his 1676 book *Musick's Monument; Or, a Remembrancer of the Best Practical Musick, Both Divine, and Civil, That Has Ever Been Known, to Have Been in the World,* Thomas Mace described a process used by lute makers in the construction of their instruments which involved the use of 'little pieces of Paper, so big as pence or two-pences, wet with Glew'. In 1845, William Shecut and Horace Day of New York filed a patent for a 'mode of preparing adhesive and strengthening plasters of india-rubber and other materials for medicinal purposes'. Shecut and Day sold the rights to their adhesive plasters to Thomas Allcock, who promoted the product as a cure for lumbago and other pains. In 1887, Johnson & Johnson began selling adhesive plasters made with zinc oxide under the ZONAS brand. Within a few years, Johnson & Johnson had observed that these adhesive plasters had 'a great variety of uses independent of its surgical application. In the household, the workshop, the factory, and in travelling, it has an almost inexhaustible list of everyday uses.' These uses included repairing glass bottles and jars, labelling containers, and mending books (although the potential of the sticking plaster was most famously exploited by *Coronation Street*'s Jack Duckworth and his impromptu spectacle repair work). Despite its versatility, medical adhesive tape had not been designed to be used in this way, and it would be several years before the first successful multi-purpose tape would be developed.

The man who invented it was Dick Drew. Drew joined the American manufacturing company 3M straight out of college in 1921 and initially worked in the research lab testing 'WetOrDry' waterproof sandpaper for quality. As part of his role, he would take the sandpaper to local car workshops to see how it coped under real conditions. In the early 1920s, a fashion emerged among car owners for two-tone paintwork. This style required a clean line between the two colours: so one area would be masked off as the other was being painted. Sheets of newspaper would be used to mask each area off, with the newspaper either glued to the body of the car or fixed using surgical tape. Unfortunately, as these pieces of newspaper or tape were peeled off, they would often damage the paintwork underneath. While delivering sandpaper to one workshop, Drew watched a painter carefully spray paint a section of a car. As the painter removed the masking, a strip of the fresh paint was also removed around the car body to the vocal frustration of the car worker. Drew told him that he could produce a better tape, despite the fact that, according to one 3M historian, 'he could back this promise with neither experience nor know-how. He didn't even know exactly what was needed, but he had the optimism of youth.'

Drew went back to the 3M offices and began work on the new tape. Even though this wasn't directly related to his job at the company, the supportive culture at 3M allowed him to continue his experiments. It needed to be easy to apply, and equally easy to remove. It also had to be sufficiently moisture-resistant to prevent paint from seeping through to the car body beneath. He based his research on the waterproof adhesive used for the WetOrDry sandpaper 3M already manufactured. This adhesive used vegetable oil and, although it was easy to remove, it stained the paint underneath. Within months, Drew's team produced an adhesive that combined a type of wood glue with glycerine; this new masking tape was easy to remove and didn't cause damage to the paint, but the kraft paper backing couldn't stretch and was therefore unsuitable for use on the curved body of a car. Also, once made into a roll, the tape would stick to itself: as it was unravelled, the glue would often stick to the layer of tape underneath. Drew tried using a cheesecloth liner to prevent this, but this made the tape costly to manufacture.

Drew's masking tape was sold in rolls two inches wide, but the adhesive was only applied to the edges (one edge would be stuck to the car body, and the other to the sheet of newspaper). According to 3M legend, one car painter complained about how tight the company was being by only applying glue to the edges and asked 'Why be so Scotch with the adhesive?' (a comment that manages to be culturally offensive to Scottish people in two different ways simultaneously; not only implying that Scottish people are miserly but also that they are called 'Scotch'). Despite its inherent xenophobia and grammatical inaccuracy, the 'Scotch' name stuck and soon became a 3M trademark. However, Drew still wasn't satisfied with the backing paper used for his tape. That was until one day he experimented with crepe paper; the crimping of the paper meant that the tape could stretch as required, yet it didn't stick to itself like the earlier tapes. In 1926, the new tape produced $165,000 in sales (equivalent to around $2.2m today). Following several improvements to the formula over the next ten years, by 1935 this had reached $1.15m annually ($19.6m today).

In the late 1920s, the American chemical company Du Pont introduced a transparent cellophane packaging material to the US. While 'somewhere between a thought and a daydream',

Drew developed the idea of using this material as the backing for a new type of adhesive tape. Four years after the initial launch of his masking tape, Drew was approached by a company which produced insulating material. The Flaxlinum Company mainly produced house insulation but in 1929 they received an order to insulate railroad refrigerator cars and needed something that could produce a completely waterproof seal. They had hoped that the Scotch masking tape would provide a solution, but its crepe paper backing couldn't completely protect the insulation from moisture. Drew experimented with different backing papers, but couldn't produce a satisfactory solution.

His masking tape, however, continued to sell well. In order to protect the product during shipping, a colleague at 3M suggested using cellophane to wrap the masking tape. If cellophane was sufficiently moisture-proof to protect the masking tape, perhaps it could also be used as a backing for the tape itself. There were still complications though, as it was difficult to apply an even layer of adhesive to the tape. Also, once applied, the amber-coloured glue would look dirty and muddy on the transparent cellophane. After nearly a year, 3M managed to produce a transparent adhesive tape and, within a decade, 3M's adhesive tape department was producing sales of $14m a year ($230m today). Although the product had been designed for use by industry and trade, sales had been boosted by the development of a smaller roll for home use. The timing could not have been better; the new product was launched in the middle of the Great Depression. This might not seem like the best time to launch a new commercial proposition, but the tape was ideal for mending torn books or other household items and appealed to thrifty consumers.

There was one problem with the product, though. Despite the improved backing materials and adhesives used for producing the tape, it still wasn't very convenient to use. Once a length of tape had been cut, the loose end would quickly disappear back into the roll. As Steven Connor explains in *Paraphernalia: The Curious Lives of Magical Things*, to locate the end of the tape, one has to use a fingernail 'as a kind of gramophone stylus, listening as much as feeling for the delicious, decisive little notch that will let you back in to the locked problem of the tape'. Even once the loose end had been located, a length of tape would need to be pulled from the roll, and cut using a pair of scissors. While still an issue today, this was even more problematic in Drew's day as he tried to convince people of the value of his new product, and had to work with brittle cellophanes and less sophisticated adhesives.

3M had made an attempt to produce a tape dispenser which could hold the roll of tape in place as a length was unwound and then cut, but it was still difficult to locate the loose end of tape and cutting a length of tape still required the use of scissors. In 1932, John Borden (sales manager for the cellophane tape department) developed a tape dispenser with a built-in blade. The shape of the blade meant that the loose end of the tape was held in place for the next use. The dispenser was then refined by Jean Otis Reinecke. Reinecke was an industrial designer who had previously taught at the New Bauhaus school in Chicago and would continue to produce tape dispenser designs for 3M for over twenty years. Introduced in 1961, the Décor Dispenser Model C-15, with its swooping curves and soft, pebble-shaped footprint, is still in production today. However, Reinecke's most famous design is the plastic 'snail'-shaped dispenser. Made from just two pieces of plastic, Reinecke's dispenser was cheap enough to produce that a disposable version could be given away with every roll of tape sold.

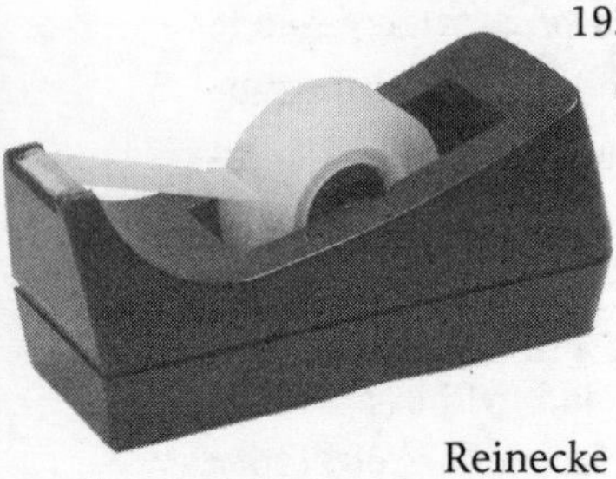

In the UK, the pressure-sensitive adhesive tape market is dominated by one brand: Sellotape. The company was formed in Acton, West London in 1937 by Colin Kininmonth and

George Gray. Kininmonth and Gray used natural rubber resin to coat cellophane film, having acquired the rights from a French company who had developed the process. As 'cellophane' was trademarked at the time, the C was changed to S, to produce the Sellotape brand name. The tape was used to seal ration and ammunition boxes during the war, and a sheet version of the product was developed to protect windows and minimise bomb damage. Just as Drew's tape had found a market during the Great Depression, the versatility of the product and its ability to easily fix household items meant it quickly became popular in post-war Britain. In a neat twist, just as Johnson & Johnson had promoted the versatility of their ZONAS medical tape by explaining how many other applications it had around the house, Sellotape would promote their tape as 'the modern bandage fix – it's quick, clean, hygienic'. In the 1960s the company was acquired by the Dickinson Robinson Group, a British packaging conglomerate; in 1980, Sellotape entered the *Oxford English Dictionary*:

> *noun*
> [*mass noun*] *trademark*
> transparent adhesive tape.
>
> verb
> (**sellotape**) [*with object and adverbial*]
> fasten or stick with transparent adhesive tape:
>
> *there was a note sellotaped to my door*

Despite Sellotape's versatility, almost half of the product's annual sales are made during the three months leading up to Christmas. During that period, the company sells around 369,000 km of tape each year as people carefully wrap presents for family and friends before the big day when the presents are ripped open and then quietly returned to the shops a few days later to be exchanged for something else. While tape is perfect for wrapping Christmas presents for ungrateful relatives, there are some jobs for which it isn't really suitable. Hanging a poster on a wall for instance – tape could damage both the wall and the poster. A pin would be even more damaging. For this you need something which is easy to remove and doesn't leave a mark.

'The ORIGINAL re-usable adhesive' it says on Blu-Tack's packaging. 'Clean and safe. Won't dry out. 1000s of uses.' Thousands of uses? I can think of maybe four. Sticking pictures to the wall; preventing an ornament from sliding off a shelf; forming a sort of protective cushion for when you make a hole in a piece of cardboard with the point of a pencil; impromptu office-based sculpture. Examining the pack more closely reveals some extra uses:

> Blu-Tack holds up: Posters, cards, paintings, decorations, maps, messages and much more
>
> Blu-Tack holds down: Ornaments, telephones, photographs in albums, screws to screwdrivers, model parts during construction or painting
>
> Blu-Tack: Cleans fluff from fabric and dirt from keyboards.

One question this brief blurb raises is how exactly do you define an individual use? Holding up posters, cards, paintings, decorations, maps, messages 'and much more' is really just one use. I studied Bostik's web site, and their marketing literature, and eventually was able to put together a list of thirty-nine uses for their product. I wrote to the company requesting a list of 1,961 additional uses for Blu-Tack.

A couple of days later, I received a reply explaining that the '1000s of uses' tagline had been used since 2005 and including a list of some of the more unusual ways the product could be used:

- The University of Cambridge contacted us asking about the softness because they hold insects' feet down with it.
- We also had an enquiry from a professor of ear surgery at

Leicester hospital saying he told children to use it as earplugs after surgery because it was the most effective method. I also saw recently in the press about using it in actual ear operations. So just for ears alone, it tends to be multifunctional.
- We have worked with the police (they approached us) about promoting the use of Blu-Tack® for holding Sat Navs in place so there are no marks on windscreens for thieves to see.
- We are also asked many times about different colours than blue and someone asked us for flesh-coloured tack because she used it in her first aid course to hold objects on their dummy in class.

Despite Blu-Tack's name, Bostik have occasionally produced differently coloured adhesive tack (including a yellow version for Marie Curie Cancer Care, and a pink version for the Breast Cancer Campaign). When Blu-Tack was originally developed, it was white in colour, but the blue colouring was added after concerns were raised that children might think it was chewing gum and attempt to eat it. Since its development in 1969, the blue adhesive putty has become a household institution with around a hundred tonnes produced at Bostik's factory in Leicester each week.

It seems that Blu-Tack was actually invented by accident. During an attempt to develop a new sealant from chalk powder, rubber and oil, an experimental reject was later discovered to have useful properties. Quite who was responsible for the failed experiment and the discovery of its handy by-product remains a mystery even to Bostik. In 2010, a story appeared in the *Leicester Mercury* marking the fortieth anniversary of Blu-Tack. 'Blu-Dunnit?' asked the *Mercury*. 'It is one of Leicester's most famous exports – but no one seems quite sure who invented Blu-Tack. And as manufacturer Bostik prepares to celebrate 40 years of the sticky putty, an appeal is being made for the mastermind to step forward.'

The Wikipedia page for Blu-Tack credits Alan Holloway of

the Hampshire-based sealant company Ralli Bondite with the initial discovery of this product and suggests that as there was no real commercial value attached to the putty, it was happily shown to any clients visiting the offices. The name of Alan Holloway was added to the Wikipedia page in November 2007 (although no citation is given). There was indeed a sealant company based in Waterlooville, Hampshire called Ralli Bondite but this was dissolved in 1995 and so employee records are difficult to trace. The Wikipedia user (Coltrane67) who made the original changes has never edited any other Wikipedia page before or since. Is Coltrane67 actually Alan Holloway (or a close relative)? It's impossible to know. (If you are Coltrane67 and/or Alan Holloway then please do get in touch.)

Although Blu-Tack is the best known adhesive tack in the UK, there are lots of similar products produced by other companies around the world. In 1994, Sellotape was taken to court by Bostik, who claimed that Sellotape's blue Sellotak product infringed their intellectual property. Although the Sellotak packaging featured Sellotape's yellow ribbon motif, Bostik believed that the blue colouring of the adhesive putty could be confused with Blu-Tack once the packaging had been removed. Bostik's argument was unsuccessful and it was ruled that as the colour of Sellotak could not be seen without removing the packaging, it was not possible that the two products could be confused at the point of sale. But, despite Sellotape's victory in the courts, the product flopped. In the US, Elmer's (an adhesive manufacturer based in Ohio) produces a terracotta adhesive putty called Elmer's Tack. UHU produce a white putty known as Patafix throughout Europe. In the UK, Patafix is sold as White Tack and on its packaging UHU claim it has 'thousands of uses'.

I contacted UHU asking for a list of uses and received the following response from a company representative:

> I just tried out in Google to type in '1000s of uses' and actually got 180,000 hits of site of products/services with

> 1000s of uses. I think this is a general English term which is used to describe a product which is really versatile in terms of usage as this is also the case for our UHU White tack/UHU Patafix.

The company apparently ignored the fact that there is already a word to describe a product that is really versatile – 'versatile'.

One use that adhesive tack does have beyond simply sticking up posters is as an art material. In 2007, Wimbledon-based artist Liz Thompson created a sculpture of a spider using four thousand packets of Blu-Tack. The 200 kg sculpture was displayed as part of an exhibition at ZSL London Zoo. Slightly less elaborate than Thompson's effort but easier to re-create at home is Martin Creed's *Work №.79* from 1993, described on the artist's web site as 'Some Blu-Tack kneaded, rolled into a ball, and depressed against a wall. Approximately 1 in/2.5 cm diameter.' According to *Frieze* magazine, 'the coloured adhesive matter alludes to the wall's function as a support and is itself supported by the wall'. The *Sun* was not quite as impressed, claiming that 'Turner art prize judges Blu it with tacky choice' after Creed was awarded the famous prize in 2001.

The *Sun* has even taken issue with more prosaic uses of Blu-Tack. In 2012, the UK government's Health & Safety Executive issued a statement on the 'Myth Busters' section of their web site in response to claims that a school in Perth & Kinross had been told by the private company which manages the property that they 'cannot display children's work on windows using Blu-Tack due to health and safety concerns' as 'a chemical in the Blu-Tack may combine with a chemical in the glass to make it shatter.' The Health & Safety Executive statement concluded that 'whatever the reason for banning the use of Blu-Tack, it is not on health and safety grounds. The manufacturer's web site makes clear that the product can be used on glass. We see no reason why the children's creative work should not be displayed for everyone to enjoy!' However, despite this statement from the Health & Safety Executive, the *Sun* still claimed that a teacher was 'banned by elf and safety ninnies from using Blu-Tack on classroom windows – in case they EXPLODED!'

After several months of correspondence with Michelle, the product manager for Blu-Tack at Bostik, in an attempt to get them to send me a list of the 'thousands of uses' their packaging promised, she sent me an email saying:

> The thing is – Blu-Tack is all about fun, creativity and imagination – this is what we say about the brand. Now if we were to come up with 1000s of suggestions for people then it would kind of take a little bit of that magic away.

I'd never really thought about the 'magic' of Blu-Tack before. Michelle sent through a list of about two hundred and fifty uses of it (compiled with assistance from her colleagues from Bostik South Africa and Bostik Australia). I didn't read it. I wanted to keep the magic alive. She also sent me a free pack of Blu-Tack, which I put in my desk drawer for safe keeping. I will never open it. It's a special memento of my correspondence with Michelle. That is one use for Blu-Tack which doesn't appear on any list.

Hypertext on a refrigerator door

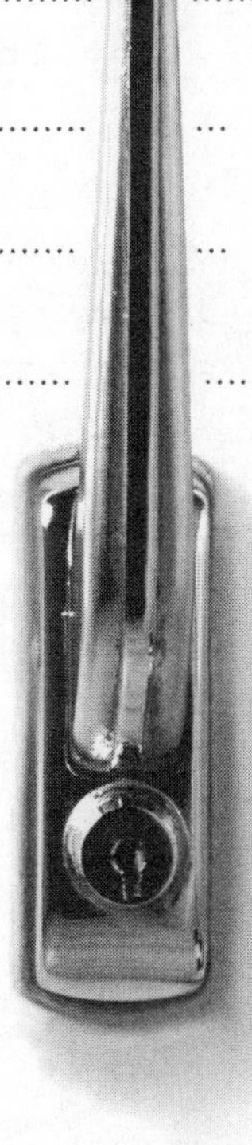

It's not revealing any major surprises to say that in the 1997 film *Romy and Michelle's High School Reunion*, starring Lisa Kudrow and Mira Sorvino, Romy and Michelle attend their high school reunion. As they travel back to their hometown, they realise that perhaps their lives weren't quite as impressive as they thought, and so they decide to reinvent themselves as successful businesswomen. Romy suggests that the pair should say they own their own company, selling a product they invented:

> I think it should be, like, something that everybody has heard about but nobody really knows who invented it. Oh my god, I've got it – Post-its! Everybody knows what Post-its are!

'Yeah!' Michelle replies, 'They're the little yellow things with the Stickum on the back, right?'

Unfortunately, they don't succeed in convincing their classmates that they invented Post-its, but they do learn that they have each other and that friendship is more important than worrying about what other people think of you and it all works out in the end because Alan Cumming invented some sort of rubber. Something like that, anyway.

But if the fictional characters Romy and Michelle didn't invent the Post-it Note, then who did?

Spence Silver joined 3M in 1966 as a senior chemist in the company's research laboratory. Silver had studied chemistry at Arizona State University before completing a PhD in organic chemistry at the University of Colorado. The team he joined were working on developing pressure-sensitive adhesives. To work effectively, these adhesives needed to be tacky enough to stick to the surfaces being joined together, but also needed to be easy to peel apart (as Dick Drew discovered, a roll of tape which stuck to itself wouldn't be particularly useful). During one experiment, part of 3M's 1968 'Polymers for Adhesives' research programme, Silver changed the formula of the adhesive he was working on, as he later explained to the *Financial Times*:

> I added more than the recommended amount of the chemical reactant that causes the molecules to polymerise. The result was quite astonishing. Instead of dissolving, the small particles that were produced dispersed in solvents. That was really novel and I began experimenting further.

The particles formed tiny solid spheres which (because of their shape) only made contact with a small part of the surface they were applied to. The result was a very weak adhesive; for a company interested in producing strong glues, this wasn't very useful. The new adhesive was also 'non-selective', meaning that if it was used to stick two surfaces together, and they were then pulled apart, sometimes the adhesive would stick to one surface and sometimes it would stick to the other. It was unpredictable. Silver was fascinated by this new substance, and was sure it could be used for something, he just didn't know what.

3M was founded in 1902. Originally known as the Minnesota Mining & Manufacturing company, it formed after a mining prospector named Ed Lewis discovered what he believed to be a deposit of corundum in Duluth, Minnesota. A form of aluminium oxide, corundum was rapidly becoming more valuable as the tough mineral was used as an abrasive in manufacturing. Five local businessmen (Henry Bryan, Dr J. Danley Budd, Herman Cable, William McGonagle and John Dwan) formed the company hoping to capitalise on Lewis's discovery.

Unfortunately, there were two major problems with this plan. While they had been planning how to transform their corundum into grinding wheels and sandpaper, a man named Edward Acheson had developed an artificial substitute for corundum known as carborundum, which caused the value of the mineral to plummet dramatically. Secondly, it turned out that Lewis had been wrong all along. It wasn't even corundum he'd discovered; it was low-grade anorthosite – similar in appearance but not strong enough to be used as an abrasive.

Not realising that their investment in the Duluth mines was a dud, the five men began construction of a large manufacturing plant to produce sandpaper from the mined material. Difficulties in mining the 'corundum' led them to begin using garnet instead. Unable to find a reliable domestic supplier of garnet, the company was forced to import an inferior form of the material from Spain. In 1914, 3M began to receive complaints that the abrasive material would fall off its sandpaper after being used for just a few minutes. No one could understand what was causing this problem, until examination of the garnet revealed that it contained some kind of oil, which made it completely unusable. An investigation revealed that a recent shipment of the material had been brought from Spain in a steam ship, which had also been carrying barrels of olive oil. During the rough sea crossing, some of the barrels had become damaged and leaked into the garnet. This experience demonstrated to 3M that they needed some way of guaranteeing the quality of the materials they were using, and so in 1916, the company built its first research laboratory, and gradually began to devote more attention to researching new forms of adhesives to improve their products.

In 1921, an ink manufacturer named Francis Okie wrote to 3M asking for samples of the sandpaper grit used by the company as he was developing a new product. 3M were reluctant to send any samples to a potential rival, but were intrigued by his letter and invited him in for a meeting. Okie explained that he had developed a new method to produce waterproof sandpaper and, rather than sell him the grit, 3M instead bought his invention and asked him to join the company to continue working on it in their new lab. Okie's WetOrDry sandpaper would go on to become the company's first successful product and inadvertently helped to move the company into the adhesive business.

As Romy and Michelle drive to Arizona to attend their high school reunion, they begin to work on their explanation of how they invented the Post-it Note. Romy imagines the pair of them as advertising executives working on a presentation for a client. As they were working on their pitch, they realised that they had run out of paperclips. 'OK,' Romy says to Michelle, 'I say, wouldn't it be great if there was, like, this, like, Stickum on the back of this paper, like, so it would – like, if I laid it on top of that other paper, it would just stay, like, without a paperclip?' Michelle seems excited as Romy adds more details, 'So then you've got like this grandfather or this uncle that has, like, a paper company or paper mill and he's like really into it and the rest is history.' In Romy's imagination, the Post-it Note was invented by a simple process: a problem was identified (Romy and Michelle had run out of paperclips) and a solution was found (they stuck some glue on the back of a bit of paper). In reality, the process of inventing the Post-it Note was the opposite of this simple scenario. As Spence Silver would later write, 'My discovery was a solution waiting for a problem to solve.'

Following his accidental invention, Silver spent several years experimenting with different formulas and playing around with different ideas, trying to find an application for this unique creation. He showed it to his colleagues, and even held seminars to explain its unusual properties. Initially, he thought the adhesive could be sold in an aerosol form – to be sprayed on the back of a sheet of paper or poster to create a temporary display. Alternatively, he wondered if it would be possible to create large notice boards coated in the material, to which memos or notes could be temporarily attached. However, the 'non-selective' property of the adhesive meant its usefulness remained limited – sticking a poster up could result in a sticky patch being left behind on the wall once the poster was removed.

One of the 3M employees who attended Spence Silver's seminars on his adhesive was Art Fry. Fry worked in the company's Tape Division and part of his role involved developing new product ideas. In his spare time, Fry was a keen member of his local choir, and a couple of evenings after hearing Silver describing his discovery, Fry found himself becoming frustrated during hymn practice. The pieces of paper he used to mark the pages in his hymn book kept falling out. If only there were some sort of low-tack adhesive he could use to hold the bookmarks in place. He went back to Spence to get a sample of the adhesive and applied a narrow strip of it to a small piece of paper to use as a bookmark. It worked, but it left a sticky residue on the page. Eventually Fry managed to develop a chemical primer that could be applied to the paper before the adhesive was added, preventing it from sticking to the page when the bookmark was removed. He showed his bookmark to his colleagues but they weren't particularly impressed. One day, Fry was in his office preparing a report. He wanted to write a brief note for his supervisor so took one of his bookmarks and jotted down a few words on it and stuck it on the front of the report. His supervisor took another of Fry's bookmarks and stuck it next to a paragraph that needed correcting, adding a few comments of his own. Seeing this, Fry had a 'eureka, head-flapping moment' and the sticky note was born.

Given how the company had only survived by ingeniously overcoming a series of setbacks (corundum which wasn't corundum, the development of artificial abrasives by Edward Acheson, oily garnet) and then found its first success as an adhesive manufacturer rather than as a mining company, it's not surprising that 3M has a strong culture of innovation. It was this that enabled Richard Drew to develop his tape when he was supposed to be producing sandpaper, and which allowed Silver to devote so much time to his useless glue. The '15 Percent Rule' at the company means that staff can devote a certain amount of their time to projects outside of their normal assignments. The belief is that this creative freedom will allow them to make discoveries they wouldn't otherwise make if they were too focused on meeting deadlines and hitting targets and encourages collaboration between members of different departments and between people with different areas of expertise. 'At 3M we're a bunch of ideas. We never throw an idea away because you never know when someone else may need it,' Fry would later observe. In the case of the Post-it, this eventually resulted in a viable product, but even once Fry had discovered a use for Silver's adhesive with his sticky note idea, it was still difficult to find support for the product within the company.

Convinced his idea would work, Fry set about building a machine which could manufacture the sticky notes in his basement. He produced a working prototype, but unfortunately, it was too big to fit through the door. Fry removed the door, then the door frame, and then part of the wall and eventually managed to get the machine out of the house and into the 3M lab. He could now produce the samples of the new product which were essential to the success of the concept. Without the samples, the sticky note would never have succeeded.

The problem was that it was difficult to convince people that the product was of any practical benefit. If you've never seen or used a Post-it Note before, then the idea must seem quite pointless. A small bit of paper with a thin strip of weak glue along one edge doesn't sound particularly useful. But once you start using it, everything changes. Fortunately, Fry's boss, Geoff Nicholson, believed in the sticky note and encouraged him to continue working on it. Nicholson began to hand out samples of the product to people in various 3M departments. It was common practice to give out samples throughout 3M and they were always gratefully received (everyone likes free stuff), but the reaction this time was different – Nicholson's secretary began to get swamped by requests for more samples. Despite this, 3M's marketing director still wasn't convinced by its commercial value. Would people really pay for a product like this when they could just use scrap paper? Inundated with requests for more sticky notes, Nicholson's secretary snapped. 'Do you want me to be your secretary or your distributor?' she asked him. Nicholson told her to forward all requests for new samples to the marketing director, who was soon equally inundated and was forced to accept that the product had potential.

Unfortunately, during the product's initial trial in 1977, consumers were as sceptical as the 3M marketing department had originally been. The Press 'n Peel Note (as it was called then) was trialled in four cities, but failed in all of them. Nicholson travelled to one test market to see what the problem was. Again, it seemed that people needed to sample the product before they bought it. In 1978, and with the backing of CEO Lew Lehr, a team from 3M descended on the town of Boise, Idaho and handed out countless samples in an exercise known within the company as the 'Boise Blitz'. Ninety per cent of those who tried it said that they would be willing to buy the newly renamed Post-it Note. This gave 3M more confidence in the product, and it was finally launched nationwide with a full advertising campaign in 1980.

The slightly hesitant launch of the Post-it Note caused problems for the producers of *Million Dollar Money Drop* in 2010, when they asked contestants Gabe Okoye and Brittany Mayt to name which product was sold first – the Sony Walkman, the Apple Macintosh or the Post-it Note. Okoye and Mayt chose the Post-it but were told they were wrong (the Walkman was introduced in 1979). Following an outcry online, the show's producers invited the couple back on the show, although sadly the series was cancelled before they had a chance to reappear on the programme.

After its nationwide introduction, and twelve years after Silver's original discovery, the Post-it was an enormous success, with Fry and Silver later inducted into the 3M Hall of Fame. The Post-it range now includes sixteen different types of products (including Page Markers, Bulletin Boards, and Easel Boards) in dozens of colours. When Romy described to Michelle her vision of how the two of them had invented the Post-it, Michelle was slightly offended. Romy appeared to be taking all of the credit. 'OK, you know, so we could say that you were, like, the designer. Like, I thought of them but you thought of making them yellow,' Romy offers by way of compromise. In fact, the reason that the Post-it was originally yellow wasn't the result of a conscious design decision. Like the invention of the note itself, this too was an accident. 'There just happened to be some yellow scrap paper in the lab,' Nicholson would later explain to the *Guardian*.

Although other companies quickly launched their own versions of 3M's product (my favourite being the Switch Note by SUCK UK – a sticky note with a hole in the middle to fit over a light switch), the Post-it Note remains iconic. In *Sex and the City*, when Carrie Bradshaw meets up with Miranda, Charlotte and Samantha to tell them that her boyfriend has broken up with her, she doesn't say 'Berger broke up with me on a sticky note'; she says he broke up with her 'on a Post-it'. Romy and Michelle would have impressed no one if they'd claimed to have invented the 'sticky note', it had to be the Post-it. Like the Pritt Stick or Sellotape, the Post-it has become not only the generic term, but also the definitive term for its type.

The appeal of the Post-it Note is simple. It helps us remember. Stuck to a finance report or on a cupboard door, the Post-it is equally as useful in the office or at home. A visual prompt, reminding us to buy milk or to send that email. In his book *The Seven Sins of Memory: How the Mind Forgets and Remembers*, Daniel L. Schacter quotes US memory champion Tatiana Cooley who says that she is fairly absent-minded in her day-to-day life. 'I live by Post-its,' she admits.

The flexibility of the Post-it, endlessly repositionable, means that it is often used by authors attempting to map out a story. Describing his writing process in the *Guardian* in 2007, Will Self explained that his books 'begin life in notebooks, then they move on to Post-it Notes, the Post-its go up on the walls of the room'. When finished, Self then moves the Post-its down from the wall and keeps them in scrap books ('I can't throw anything away'). Reflecting on this flexibility, and the way it allows information to be linked together, Paola Antonelli, curator of architecture and design at New York's Museum of Modern Art, described the Post-it as 'hypertext on a refrigerator door' when she included it in her *Humble Masterpieces* exhibition in 2004.

After the struggle to get people to understand the usefulness of the Post-it, the sticky note remains a familiar sight even on our computer desktops. The 'comment' facility in Microsoft Excel appears as a small yellow square, but other applications make their relationship to the sticky note even more explicit – not only 3M's own Post-it Digital Notes, but also Apple's Stickies and Microsoft's Sticky Notes. However, the most common meeting of technology and sticky note is also the simplest – a Post-it stuck to the side of a monitor.

Although it's not fair to suggest that the Post-it lives a purely functional existence. It can sometimes become art. In 2001, California artist Rebecca Murtaugh used thousands of Post-it Notes to cover every surface in her bedroom as part of her installation *To Mark a Significant Space in the Bedroom #1*. Different values were given to each Post-it colour; the original canary yellow was used for low-value areas such as the walls and ceilings; brighter neon colours were used for her favoured possessions. Commenting on the installation for the *New York Times*, Murtaugh explained that she was 'mesmerised' by Post-it Notes.

> They come in all these colors. They are beautiful. They have a purpose, but it's different for everyone: sometimes it's a note, 'I'll be back,' or maybe it's a phone number. But for all these important things, the note itself is always ephemeral and temporary. Yet it's carrying all this valuable information. So there is this duality: it's disposable, but it's very valuable. I wanted to mark an important space, not a book with a note, but an entire room.

Murtaugh has continued to use Post-its in her art, although no longer uses them for their original purpose. 'I don't want to waste them,' she says.

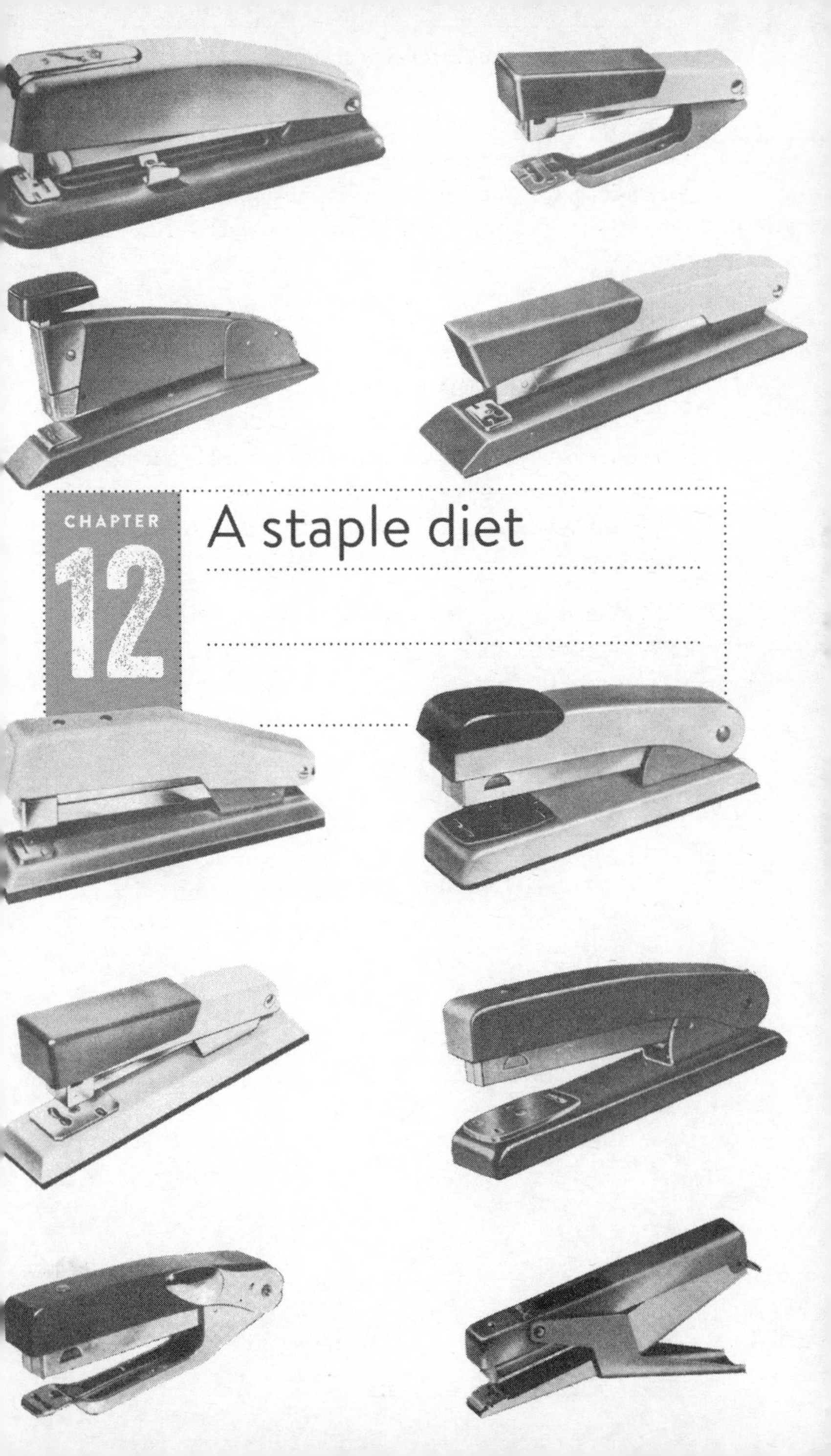

CHAPTER 12

A staple diet

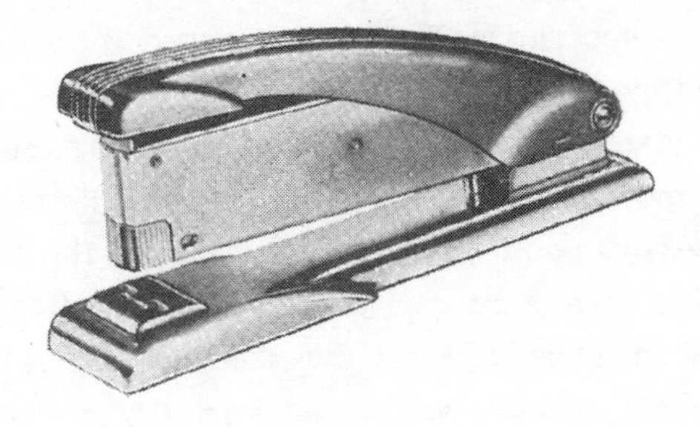

Few novels have examined the modern office environment in such intricate detail as *The Mezzanine* by Nicholson Baker. In one beautiful passage, the unnamed protagonist describes the three stages he looks forward to as he presses down on the 'brontosaural head of the stapler arm' to staple together a thick document. The first stage, as you press down on the stapler and feel 'the resistance of the spring that keeps the arm held up', followed by the second stage, as the stapler blade 'noses into the paper and begins to force the two points of the staple into and through it', and then the final stage:

> the felt crunch, like the chewing of an ice cube, as the twin tines of the staple emerge from the underside of the paper and are bent by the two troughs of the template in the stapler's base, curving inward in a crab's embrace of your memo, and finally disengaging from the machine entirely.

The protagonist then describes a disastrous scenario familiar to many. Finding, with 'your elbow locked and your breath held' that the stapler is empty. 'How could something this consistent, this incremental, betray you?' The stapler might not immediately appear to be a romantic object, but Baker's description of feeling 'betrayed' by the empty stapler illustrates the emotional attachment people feel towards these items.

They're substantial. That must be part of the reason. Metal arms and springs; more complicated than the pens and pencils surrounding them. Mechanisms not fully understood; they deserve respect. They're refilled rather than replaced, and so live longer than pretty much anything else on the desk (there must be companies that replace their PCs more often than they replace their staplers). In fact, they don't just outlive the rest of the stationery an office worker may use, they probably outlive most office workers themselves. You'll move on to another job, but your stapler will remain. Returned to the stationery cupboard, awaiting a new owner, ready to build a new relationship. Sometimes though, the bond becomes too strong and people aren't prepared to allow a simple termination of contract to change things. According to research published by Rexel in 2011, 'during the recent recession many staplers were taken home by staff made redundant because they believed them to be personal property'.

But perhaps no one has become as protective of their stapler as Milton in the 1999 comedy *Office Space*. Distressed when his employers at Initech switched from Swingline staplers to Boston, Milton (played by Stephen Root) holds on to his red Swingline:

> I kept my Swingline stapler because it didn't bind up as much, and I kept the staples for the Swingline stapler and it's not okay because if they take my stapler then I'll set the building on fire.

In fact, when the film was made, Swingline didn't actually produce a bright red stapler like the one Milton defends so vehemently. Now part of the ACCO group, Swingline was founded by Jack Linsky, who had moved to New York from Russia as a child. When he was fourteen, Linsky began working for a stationery supplies company. He started a wholesale business and travelled to Germany to visit stapler factories. He believed the designs of the machines on the market at that time could be improved and made more streamlined. Unable to convince any German manufacturers of his vision, he opened his own company, the Parrot Speed Fastener Company. The company developed a 'top loading stapling machine' which was 'practical and efficient in operation'. This new design was the first to allow users to swing open the top of the stapler to refill it or remove any bent or damaged staples. Linsky's wife, Belle, suggested the name 'Swingline' for this new type of stapler, and it was so successful that in 1956, it became the company's name.

While Swingline had produced red staplers alongside the usual blacks and greys in the past (the Tot 50 and the Cub for example), they had been discontinued years before *Office Space* production designer Edward T. McAvoy tried to track one down. Director Mike Judge wanted a bright red Swingline stapler to stand out against the drab colours of cubicle life shown on screen, but no such stapler existed any more. McAvoy was forced to improvise. He called Swingline and asked if it would be OK to spray paint one of their staplers. Fortunately for everyone involved (most fortunately of all for Swingline themselves), the company agreed. McAvoy took a handful of staplers along to a car mechanic, and had them sprayed cherry-red.

Although not a huge hit on its initial release, the film has gradually acquired cult status. Fans would ask actor Stephen Root to sign their staplers, and began to produce their own spray-painted versions. Some started to contact Swingline to ask where they could buy red staplers like the one they'd seen in the film.

Eventually, in 2002, Swingline launched the 747 Rio in bright cherry-red. 'We've been in the business for over 75 years, and this is certainly some of the biggest direct interest we've ever seen,' Swingline president Bruce Neapole told the

Wall Street Journal shortly after the new product was launched. Following the success of the 747 Rio, Swingline, who had spent most of the previous few decades trying to secure bulk orders with large companies in safe black and platinum coloured offerings, suddenly saw a market in what they termed 'expressive consumers'. The Swingline web site now invites these expressive consumers to send in photos of themselves with their staplers in unusual locations. 'Share your love' the web site says, alongside pictures of red staplers up trees, on scooters and next to hot tubs (this last one is shown sipping a cocktail, meaning I find myself in the unexpected situation of being jealous of a stapler).

Swingline's introduction of the top-loading stapler in 1939 was a crucial step in the evolution of the modern stapler, 'allowing an office worker to simply drop in a row of staples'. But the top-loading stapler we take for granted today was only possible due to a change in the way staples themselves are produced, in particular the introduction of the gummed or 'frozen' strip of staples. It is this which offers some sense of consolation to Nicholson Baker's protagonist after the 'betrayal' of his empty stapler:

> laying bare the stapler arm and dropping a long zithering row of staples into place; and later, on the phone, you get to toy with the piece of the staples you couldn't fit into the stapler, breaking it into smaller segments, making them dangle on a hinge of glue.

The earliest staplers, however, did not accept strips of staples. In fact, they could only hold one staple at a time, and would have to be reloaded after every use. The machine patented by Albert J. Kletzker in 1868 and thought to be one of the very earliest staplers was actually described in its patent application as a 'paper clip'; but this was nothing like the round-ended Gem paperclip we use to slide over a few sheets of paper today. This was a fearsome-looking beast. The metal fastener would be inserted into position between two upward-pointing, fang-shaped guides. The sheets of paper were placed on top of these guides before a lever pushed the paper down, the teeth of the guides piercing the paper on the staple's behalf. The lever would be released and 'the points of the fastener will be bent inward with the fingers, after which a second stroke of the follower will force the points of the fasteners down tightly to the papers and then the fastening will be complete'. In this way, the device acted as a sort of upside-down version of the stapler as we'd understand it today; instead of the staple being pushed down into the paper, the paper would be pushed down on to the staple. In fact, the staple or fastener itself was entirely passive during the whole process; the guides make the holes as the paper is pressed down, and the staple legs then bent manually. The first stapler to simultaneously insert and clinch the staple was patented by Henry R. Heyl in 1877. His device worked in a similar way, but as the staple passed through the paper, its legs were bent inwards towards each other.

Impressive as Heyl's design was, the first commercially successful desktop stapler was that patented by George W. McGill in 1879 (who obviously had more success with staplers than he did with his many and varied paperclip designs). Unlike the earlier designs, McGill's machine pressed the staple down and through the paper, rather than pressing the paper down on to the staple. This allowed for the operation to be carried out in one 'continuous and instantaneous' movement, avoiding the need for any second blow and so the machine became known as 'McGill's Patent Single Stroke Press'. The Single Stroke Press was more convenient than earlier models, however the staples still needed to be inserted into the machine one at a time. Daniel Somers developed a stapler with a 'feeding slide' that could hold a magazine of staples in 1877, but the machine was not as successful as McGill's Single Stroke Press, despite its apparent superior design. Somers was the Betamax to McGill's VHS.

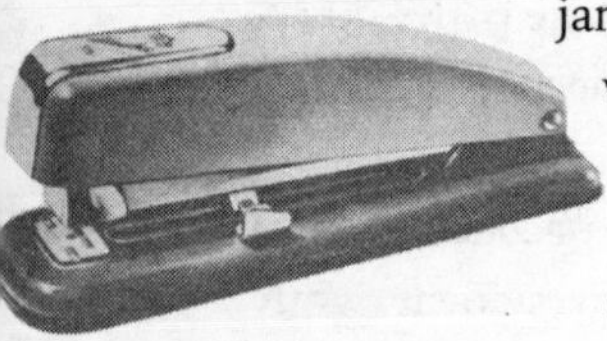

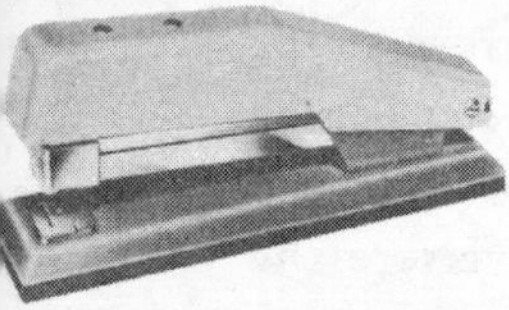

Sold on wooden or metal cores, the staples used in these early magazine-fed staplers were loose and could easily become jammed during loading. These gradually gave way to continuous strips of staples – often connected along a thin metal spine; as the stapler was pushed down, a blade would slice through the spine, severing the staple from the strip. This required a lot of force and so the metal spine was soon replaced by adhesive. Used first in the Bostich No.1 in 1924, the design of these glued staple strips has remained largely unchanged ever since. The wire of each staple is tapered very slightly so that when aligned in a strip, a minuscule mountain range is formed of peaks and troughs – these shallow channels are then filled with adhesive, holding the strip of staples together.

When refilling these early staplers with a new strip of staples, it was important to make sure the staples and stapler were compatible. Whereas modern staplers will generally accept any standard staples, there was no real standardisation for a long time, with different sizes used by different manufacturers for different models. This was inefficient for both retailers (who had to carry a wide variety of different staples) and consumers (who struggled to find the staples they needed), and so gradually standardisation crept in. The Rexel 56 range, introduced by Rexel in 1956, included a variety of different stapling machines available at different prices, but which all used the same type and size of staple. The simplicity and convenience of this system meant the 56 Range was an immediate success, with the brand still dominating the market today.

Staple sizes are given using two numbers – the gauge of the wire (usually 26 or 24 gauge) and the length of the 'shank' or leg of the staple (usually 6 mm). The most common office staple is the 26/6 (also referred to as a No.56 as this was the staple used in Rexel's 56 range). While sizes became standardised, quality still varied and so staplers would be stamped with warnings ('USE ONLY GENUINE "BRINCO" STAPLES IN THIS MACHINE. OTHER KINDS MAY BLOCK IT' or 'GUARANTEE VALID ONLY IF GENUINE REXEL JUNIOR No.46 STAPLES ARE USED'). In a classic cake-and-eat-it operation, manufacturers would try to scare customers off from using staples sold by their competitors, while at the same time trying to sell their staples to users of rival machines. The boxes of staples they sold would list the machines they were compatible with ('For use in Velos Sprite Stapler/Velos Jiffy Plier/Little Peter, Speedy Swingline Tot/Tatem Buddy Junior').

The staples used in the earliest machines, before the Bostitch glued strip, tended to be much thicker than the staples we are used to today. Removing them was almost as difficult as inserting them. But perversely, no real solution to the problem of staple removal seems to have been developed until after the use of thinner wire staple became widespread. One possible reason for this is that the early staples were so thick they required pincers to remove, and pincers had already been invented. But these thin staples almost seemed like they could be removed by hand; sliding a fingernail underneath one end. Except, as we have all discovered for ourselves, that can be painful and doesn't really work.

In 1932, William G. Pankonin of Chicago applied for a patent for his 'Tool for Removing Staples', which allowed the user to 'quickly remove staples or similar fasteners from paper without tearing or mutilating the paper'. This design was similar to a small pair of pincers; the ends of the jaws would be passed under the body of the staple, and as the stapler remover was pulled up, the legs would 'ride across the paper' on the underneath of the document, straightening out as the staple was removed. Later designs (such as that patented by Frank R. Curtiss in 1944) more closely resemble the staple remover we are familiar with today, with its winged finger grips and fanged jaws. With the fundamentals of the staple remover concept in place, the design has barely changed in decades.

Along the top of my desk I have a small collection of staplers organised in roughly chronological order, one for every decade of the last one hundred years. Looking at them, you see a compressed century of design in front of you. Things becoming flatter, thinner, shinier, curvier. You would learn nothing from a similar collection of staple removers. Staplers seem to follow fashion, but there is no real need for the staple remover to worry about such things. Staplers will often sit on your desk. You want them to be close at hand as they are used so frequently. They're visible. They need to look nice. Smaller, and less frequently used, staple removers lie in desk drawers or in cupboards. Occasionally, a novelty manufacturer might try to emphasise the fact of the remover's fangs by fashioning it into the shape of a crocodile or snake's head, but such things have no place in a serious office.

One way to avoid the need for staple removers altogether is to pin rather than staple your documents together. Generally, the legs of a staple bend inwards; this is the most secure way to hold documents together. By rotating the anvil face (there's usually a little button on the underneath of the stapler), the legs can be made to point outwards, almost creating a straight pin which can easily be removed by hand. For some, however, two options (legs pointing inwards or legs pointing outwards) are not enough. As well as developing a staple remover, William Pankonin also developed a stapler anvil which gave much more choice to the user. His 1934 'Anvil for Stapling Devices' would sit in place of the standard two-way anvil and consisted of two separate dies or 'seats', one for each staple leg. These dies could be rotated independently of each other into a number of different positions. The dies could be aligned with one pointing inwards and one pointing outwards, creating a 'hook-ended staple'; or one could point towards the front of the machine and one towards the back, creating a 'Z-shaped staple'.

Alternatively, they could be aligned pointing inwards (as with a conventional stapler) or pointing outwards (resulting in what he describes as 'an elongated member which may be readily withdrawn', offering the closest that 1930s stapler patent applications come to sounding like dialogue from a *Carry On* film). While the flexibility of this design is very impressive, it seemed to struggle against the fact that it is almost entirely pointless.

While standard 26/6 or 24/6 desktop staplers are suitable for most jobs, there are times when you want something a little more hardcore. You need a bit more power. Early electric staplers, such as the Bostitch Electromagnetic Fastener Model 4 (patented in 1937) simply consisted of a metal arm attached to a motor that would press down on 'a standard type of desk stapler' and were controlled by a foot pedal. The 1956 Bostomatic also featured a standard Bostitch desk stapler, but would operate automatically when papers to be stapled were inserted against the 'featherweight touch switch'. Modern electric staplers designed for office or domestic use can staple up to seventy sheets of paper and hold over 5,000 staples in specially designed cartridges.

As a sort of compromise between the electric stapler and the standard hand-operated model, there are the 'reduced effort' staplers, like the PaperPro Prodigy. Using a spring mechanism, the PaperPro is able to insert staples without as much force as a standard machine (the company claims the Prodigy can staple through twenty sheets of paper using the equivalent of 7 lbs of force, compared to 30 lbs for a standard stapler). 'You can do twenty sheets with one finger,' Todd Moses, the CEO of Accentra who make the PaperPro, told *Time* magazine in 2005. 'You can even use your pinkie.' Impressive as this sounds, not everyone is convinced. As Swingline's vice-president Jeff Ackerberg told the same magazine, 'There's a natural need to make staplers easier to use, but there's not a natural need to use a pinkie.'

No natural need to use a pinkie, but is there even a natural need to use a staple at all? Sold today as an environmentally friendly way to prevent waste, staple-less staplers bind papers together by 'cutting and folding the paper in one action without using a metal staple' (a small slit is cut through the document and this is then folded back in on itself), but in fact this concept has been around for over a hundred years. George P. Bump of La Crosse, Wisconsin applied for a patent in 1910 for his design which held sheets of paper together by 'cutting a tongue from the superimposed sheets and turning this tongue back and threading it through a slit formed in the sheets so that the said sheets will be held together by the tongues cut from the paper sheets themselves'. Where now the idea is promoted as a way of saving the planet, back then it was promoted as a way of saving your pennies ('Practice vs. Precept. The Government advocates THRIFT. The Government is also buying "BUMP" Paper Fasteners. Why? Because in Practice as well as Precept, the "Bump" Paper Fastener is the synonym of THRIFT'). It's fortunate that we no longer live in a time when we have to worry about the economy or government efficiency.

CHAPTER 13 The storeroom of knowledge

'I will not be pushed, filed, stamped, indexed, briefed, debriefed or numbered. My life is my own.'

It's one of the greatest resignations in television history. Episode One of *The Prisoner*. Patrick McGoohan marches down a long underground corridor, storms into the office of his boss, slams his resignation letter down on to the table, thumps the desk and knocks over a cup of tea (hitting the desk so hard that the saucer breaks). The thunderbolt sound effects only add to the drama. As McGoohan drives off in his yellow Lotus Seven, a series of ×'s are typed across his photograph, and an automated filing system drops his record card into a drawer labelled 'RESIGNED' in a giant room filled with filing cabinets. As impressive as this facility is, with its rows of grey metal filing cabinets continuing off into the distance, it's nothing compared to the 'Storeroom of Knowledge' in Terry Gilliam's *Brazil*. Filled with 'towering filing cabinet skyscrapers', the Storeroom contains 'every bit of knowledge, wisdom, learning, every experience, every thought neatly filed away'.

In terms of scale, the closest we have in the real world to the Storeroom of Knowledge is *Untitled (Minuet in MG)* by

Samuel Yates. Yates's 1999 sculpture consists of a seven-storey tower of filing cabinets, containing 'a 1974 MG Midget sports car that was donated, shredded, steamrolled, photographed, bagged, labelled, numbered, and filed by weight from heaviest to lightest'. Reaching a height of sixty-five feet, the sculpture holds the Guinness World Record for 'tallest file cabinet on earth'. The fifteen filing cabinets used by Yates to create the sculpture were provided by the HON Company. Originally known as Home-O-Nize (presumably this was some sort of pun on the idea of 'harmonising' the home, but a pun so bad you can only guess whether or not it was even intentional), the HON company was formed in 1944 by engineer Max Stanley and his brother-in-law, advertising executive Clem Hanson, along with industrial designer H. Wood Miller. Initially the plan was to provide work for American servicemen returning from war and Home-O-Nize sold items that were simple to manufacture, such as coasters and recipe boxes. In 1948, the company introduced a range of filing cabinets and began to focus their energy on the office supplies industry. By the early 1950s, sales had reached $1m, and today, HON (now HNI International) is the second-largest office furniture manufacturer in the world.

Despite its size, HNI International is not particularly well known in the UK. If Yates were to produce a similar sculpture in this country, it is likely that he would use Bisley filing cabinets. Named after the town of Bisley, near Woking in Surrey, the company was formed by Freddy Brown in 1931. Originally, Brown focused on car repairs, but in 1941, the company began producing metal containers for the Air Force to drop supplies by parachute and moved to a larger factory to meet demand. Once the war was over, the market for large metal containers that can be dropped by parachute dramatically declined. Fortunately the company was contacted by a wholesaler called Standard Office Supplies asking if they could produce metal waste paper bins instead. During the next few years, the company began to concentrate on office supplies and in 1963 the car repairs part of the business was dropped.

Freddy's son Tony joined the company in 1960. Freddy had five children, but it was Tony who really took to the business. When Freddy retired in 1970, it was Tony who took over, buying out the shares of his other family members for £400,000. Tony pushed the company further into the office furniture market and during the next few years, the design team (led by Bernard Richards) began to produce the simple metal filing cabinets for which the company would become best known. Ball-bearings in the sliding mechanism meant that the drawers opened and closed smoothly, and extending metal runners allowed '100% extension'. Despite selling more than any other brand of filing cabinet throughout Europe, the Bisley range remains humble. The wide range of colours allows it to camouflage itself within any office environment; muted tones for traditional offices and brighter, bolder colours for the more adventurous workplace. According to design critic Jonathan Glancey in the *Guardian*, the Bisley cabinet is 'as discreet as an English butler in the age of Jeeves and Wooster'.

Before the vertical filing system we use today was developed in the 1890s, incoming mail would be folded and stored in pigeon-holes above the office worker's desk after it had been dealt with. Each letter would be abstracted, with the date received and sender's details written on the outside of the folded letter and filed away. For the low level of correspondence that was the norm prior to the mid-nineteenth century, this system was perfectly adequate. However, a combination of things (growing industrialisation, the telegraph, the railway and postal reforms) all made it easier and cheaper to communicate and do business over long distances. This in turn led to the development of larger corporations and the associated increase in bureaucracy. Letter press copiers and aniline dyes made it quicker for outgoing correspondence to be duplicated (without letters having to be copied out by hand). As the cost of outgoing correspondence decreased, the volume increased: the pigeon-hole storage system could no longer cope.

Flat filing was developed during the second half of the nineteenth century. Letters no longer had to be folded or abstracted, which made it easier to store and retrieve them. The first step towards flat filing was to simply bind incoming mail in a book, but this was soon replaced by the box file, which 'consisted of a box, its cover opening like a book, with twenty-five or twenty-six pages or pieces of manila paper, tabbed with the letters of the alphabet and fastened into the box at one side, the papers being filed between the sheets'. As they were not bound in the file, letters could easily be rearranged, making the system much more flexible than previous filing methods. Although these horizontal flat files were more convenient than the previous pigeon-hole system, accessing a document at the bottom of a box file still meant lifting up all of the other documents on top of it. Hardly ideal.

The Shannon file, designed by James Shannon in 1877, consisted of 'a small letter-size file-drawer about the size of a loose-sheet file except that the drawer consists only of bottom and front'. It had no sides, 'sides being unnecessary on account of the arches on which the papers are filed'. In Shannon's earliest designs, the wire arches featured sharp points which would pierce the documents, but later provided 'a punch adapted by a single motion to cut clean-edged holes in the papers corresponding with the number and position of the receiving-wires' (a hole punch, in other words). Around the same time, a filing system based on a similar concept was being developed in Germany. Friedrich Soennecken patented designs for both a type of ring binder and a type of hole punch in 1886. A decade later, the lever-arch file mechanism was developed by Louis Leitz.

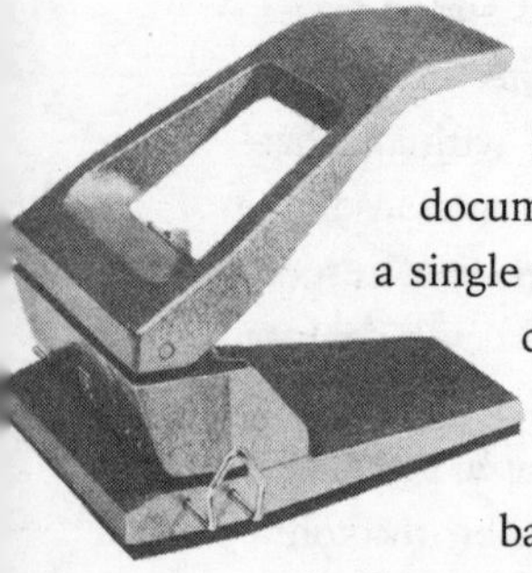

But before the hole punch, ring binder and lever-arch system could be of any real benefit, the position and spacing of the holes created would have to be standardised; if my hole punch doesn't fit your ring binder, we may as well give up. ISO 838 ('Paper – Holes for general filing purposes') specifies 'the dimensions, spacing and the position of the holes made in sheets of paper or documents in order to permit their filing in ordinary files'. According to ISO 838, 'in principle, the holes shall be arranged symmetrically in relation to the axis of the sheet or document, perpendicular to the line of the axis of the holes'. The distance between the centres of the holes should be 80 mm, the holes should have a diameter of 6 mm, and the centre of each hole should be 12 mm from the edge of the sheet of paper. It costs £26 to download a PDF of ISO 838. That's £26 to be told how to use a hole punch. I could have bought a high-end forty-sheet capacity Rapesco 835 hole punch with that money.

Additional holes can be added for extra security (again at 80 mm intervals) and because of the relationship between the sizes of sheets within the A-series, a sheet of A3 can be filed alongside A4 documents simply by punching the holes along the shorter side and folding it in half. In fact, anything from A7 and above can be filed using the same system. Inevitably, just as in the US they have resisted the appeal of the A-series, so they have also rejected the ISO 838 specifications. Instead, they use a three-hole system, which is incompatible with the rest of the world and is far less flexible. Well done America.

Edwin G. Seibels is credited with inventing the vertical filing cabinet system back in 1898. Seibels was a partner at Seibels & Ezell, a South Carolina insurance agency founded by his father. Frustrated by the inefficiency of the pigeon-hole system still in common use at that time, Seibels felt it would be quicker to keep the letters flat in larger envelopes and store them standing on end in drawers rather than folding and abstracting them. He contacted a local woodworking company, who built five wooden cabinets based on his instructions. However, Seibels would later be disappointed when he tried to patent his idea and discovered that there was nothing in it which could be protected:

> It was pointed out that by simply varying the size, a filing box could be made which would not infringe my patent. Unfortunately, I overlooked the part played in setting the envelopes upright and separating them by guide cards. This device, of course, could have been patented.

While the Seibels cabinet is very similar to the modern filing cabinet, the practice of storing documents vertically in files had already been established in the 1870s with the introduction of card index files into libraries following the development of the Dewey Decimal system. The Dewey Decimal Classification system was created by Melvil Dewey in 1876, and is familiar to library users around the world, all of whom no doubt have their own favourite classification numbers within the system (mine is 651 – Office Services). The index card catalogue was quickly adopted as a retrieval tool alongside the classification system itself; the cards were filed in cabinets with tabbed dividers allowing users to easily identify the location of a particular book or document.

The flexibility of the index card catalogue was based on a system developed by the eighteenth-century Swedish naturalist Carl Linnaeus. Linnaeus was developing a taxonomic structure of plant and animal species and was faced with two seemingly contradictory requirements: the need to put species in some kind of order, and the need to be able to integrate new species into that arrangement. His solution was to use small cards (similar in size to the 5" × 3" index cards still used today).

The way in which index cards allow information to be easily rearranged, and for new information to be added at any point, means that they are not just useful for creating catalogues or organising files, but for any creative process. 'The pattern of the thing precedes the thing,' Vladimir Nabokov told the *Paris Review* in 1967 as he described his system of working, 'I fill in the gaps of the crossword at any spot I happen to choose. These bits I write on index cards until the novel is done. My schedule is flexible, but I am rather particular about my instruments: lined Bristol cards and well sharpened, not too hard, pencils capped with erasers.'

While tabbed dividers provide a simple way of organising a set of index cards, the *whirr* of a Rolodex is much more satisfying. The Rolodex was invented by Oscar Neustadter from Brooklyn in 1950. Neustadter's Zephyr American Corporation had previously developed the Swivodex (an inkwell which doesn't spill) and the Clipodex (an aid for taking dictation which secretaries could clip to their knees), but neither of these had much commercial success. Neustadter's Autodex telephone directory did better (and is still sold to this day), but it was the Rolodex for which Neustadter will be remembered. 'I fiddled with the idea with Hildaur Neilson, my engineer,' Neustadter would recall in 1988. 'He built a model, then we started making them. I knew I had a good idea, but people were sceptical at first. The first one looked like a steel one they still make today. It had a rolltop cover. It also had a key and lock. Heh-heh. The same key fit every Rolodex in the world.'

While the Rolodex is suitable for quickly sorting through business contacts written on small cards, it probably wouldn't be suitable for use as a filing system for anything much larger (attractive as the idea might be, I don't think an A4 Rolodex would be very practical). A filing cabinet is a better solution.

The vertical filing cabinet could store more in the same amount of physical space than the older flat filing systems. One filing cabinet company produced an advert in 1909 that claimed their system was 44 per cent more efficient and would cut labour costs by a third. The introduction of the lateral suspension file, with documents grouped together in paper hammocks, increased efficiency even further.

But for many people, the files and folders they use most often are those on a computer. Although electronic filing has obvious advantages in terms of space, there is also a danger: long-term storage and retrieval becomes complicated as formats become obsolete. Data needs to be migrated from one medium to the next to prevent it becoming trapped on an out-of-date format (my A-level coursework is kept on a series of 3.5" floppy disks I can't bring myself to throw away, despite the fact that even if I got around to buying a USB floppy disk drive, the disks would probably have deteriorated and wouldn't be read anyway). In his essay on 'Very Long-Term Backup', Kevin Kelly of the Long Now Foundation compares this to the lifespan of paper documents:

> Paper, it turns out, is a very reliable backup for information. While it can burn or dissolve in water, good acid-free versions of paper are otherwise stable over the long term, cheap to warehouse and oblivious to technological change because its pages are 'eye-scanable'. No special devices needed. Well-made, well-cared-for paper can last 1,000 years easily and probably reach 2,000 without much extra trouble.

While the safest long-term storage system might therefore be to simply print out the entire contents of your computer and shove it all in a filing cabinet, it might take up quite a bit of space. One gigabyte of memory can hold around 65,000 pages of Microsoft Word, and a fairly cheap laptop will typically have at least a 500 GB hard drive – even if you printed on both sides, it's still a lot of paper, and using a 44 per cent more efficient vertical filing system won't save you either. At some point, you'll need to archive it all somewhere else.

In 1913, the US government introduced new legislation requiring businesses to keep written records for tax purposes. Previously, individual companies had their own policies regarding how long to keep documents, but the new legislation formalised record keeping throughout the United States. In Chicago, a young tailor named Harry L. Fellowes worked next door to a store owned by Walter Nickel. Nickel sold collapsible storage boxes, designed specifically for archiving written documents. Nickel was called up for military service in 1917, and Fellowes bought Nickel's stock for $50 (around $920 today). He had no trouble selling the boxes, and the business grew following the war. Back from service, Nickel rejoined the company and the pair of them expanded the product range, introducing the patriotically named Liberty Box, alongside the Bankers Box. The dark wood-grain effect Fellowes R-Kive box remains a familiar sight in offices around the world. Yet despite the fact that these boxes were sold by over a hundred companies around the world, some people were not satisfied with them.

In his 2008 documentary *Stanley Kubrick's Boxes*, Jon Ronson explored the contents of the thousands of boxes in the Kubrick archives. Rows of shelving units containing hundreds of wood-grain effect R-Kive boxes fill a huge warehouse, the boxes filled with location photographs and research materials for his films (Kubrick was a keen collector of stationery and would often visit his local branch of Ryman to stock up, and so some of the boxes are labelled simply 'green notebooks' or 'yellow index cards'). However, Kubrick had apparently become increasingly frustrated with the lids: they were too tight. Kubrick's assistant Tony Frewin contacted the Milton Keynes box manufacturer G. Ryder & Co. Ltd specifying the internal dimensions he felt were optimum for a storage box. In a memo to the company, he wrote that the lid must be 'Not too tight, not too loose but JUST PERFECT'. Ryder produced the box as follows:

Ref: R.278

Type: Brass wire stitched box, full depth lift-off lid (case lid) with triangular lugs.

Composition: 1900 micron (0.080 inch) double sided kraft container board.

Dimensions (internal): 16¼ × 11 × 3¾ inches (R.278).

In one consignment of boxes, Tony found an internal memo left by someone at G. Ryder & Co. by mistake in one of the boxes: 'Fussy customer – make sure the lids slide off properly.' 'Yeah, I guess we were fussy customers,' Tony tells Ronson, 'as opposed to the customers who didn't mind spending all afternoon struggling, trying to get a lid off.'

Once everything is safely stored in an archive box, it's essential to label the boxes clearly to have any hope of retrieving the contents. In 1935, using a motor from an old washing machine, some parts taken from a sewing machine and a sabre saw, Ray Stanton Avery built a machine to produce self-adhesive labels. Ray's future wife, schoolteacher Dorothy Durfee, invested $100 in the business, allowing Avery to form the Klum Kleen Products company (wisely, the name was changed to Avery Adhesives the following year). The product was an instant success, and by the time Ray died in 1997, annual sales at the Avery Dennison Corporation had reached $3.2bn.

Archiving and retrieving documents stored chronologically can be made easier with a simple date stamp. A couple of years ago, I went to New York and went into a shop on Second Avenue called Barton's Fabulous Stationers. 'Fabulous stationery', promised the awning outside, and I was not disappointed as I wandered around the store. In many ways, it reminded me of Fowlers back in Worcester Park where I'd bought my Velos 1377 Revolving Desk Tidy. Just like Fowlers, the shop was split – one half selling gifts and toys, the other half selling stationery. And just like Fowlers, the shop somehow managed to stay open despite some of the stock on its shelves staying there for years. Yellowing notepads and fanfold dot-matrix printer paper ('with Letr-Trim Edges') lined the shelves. I bought a Trodat 4010 date stamp.

I've always liked date stamps, probably because for many years I worked in libraries. Each morning, we would move the date on each stamp forward by one day. The stamps used for standard loans would be clamped to the heads of the barcode scanners and, with their associated trays of letters and numbers, would be the closest I would ever come to traditional moveable type printing. For other items (CDs or videos – I left before DVDs became widespread), dial stamps would be used. These tended also to be Trodat stamps, but not the mechanical self-inking 4010. Council budgets could only stretch to the cheaper manual stamps.

The graceful pirouette of the Trodat stamp head as it moves through 180° – from contact with the ink pad through to contact with the paper, imprinting its message and then repeating its journey in reverse – is something which can only really be appreciated by slowing down the action, by looking at it up close, by using it in a way it's not designed for. You reach for it from the corner of your desk to stamp an invoice, it makes a satisfying 'thwack', but you give it no more thought than that. Yet it deserves more respect. It's a thing of beauty.

Saying that though, I've never actually used the Trodat 4010 I bought from Barton's Fabulous Stationery. Although that isn't entirely through choice; when I bought it, I noticed that the box was a bit battered, but the stamp inside was in perfect condition. It was only when I got back to London and looked at it properly that I realised how long it must have been on the shelf. The date range ran from 01/01/1986 through to 31/12/1997. I'm not sure quite how far in advance Trodat produce their date stamps, but it's probably safe to assume they don't backdate them. That Trodat stamp must have been sitting on that shelf in New York since the late 1980s, and for the last fifteen years it has been functionally obsolete.

I think the date bands are replaceable, but I'll keep it the way it is.

CHAPTER 14

Tomorrow's world

It's only a slight exaggeration to say that the history of stationery is the history of human civilisation. A straight line can be drawn (using a ruler from the Indus valley) between the bitumen used to haft a piece of flint to wood to form a primitive spear and the glue in a Pritt Stick; between the pigments used to create the earliest cave paintings and the ink used in a ballpoint pen; from Egyptian papyrus to a sheet of A4; from the stylus to the pencil. In order to think – in order to create – we need to be able to write things down, to organise our thoughts. In order to do that, we need stationery.

Or rather – in order to do that, we used to need stationery. But now? Now we have computers, the internet, email, smartphones and tablets. Our ability to record our thoughts and ideas no longer involves putting pen to paper. We can bash out a quick note on our phone on the bus, and it'll be there waiting for us when we open our laptops at home. Everything can be synced and indexed and stored in the cloud and instantly retrieved on countless devices, along with anything else we've tagged alongside it. No more scrambling around trying to find the scrap of paper you scribbled something down on. No more unintelligible scrawls or pens running out or pencils breaking or ink smudging. Just a smooth, seamless, efficient future.

Whither the ballpoint? Could it be that within a few years, stationery will be no more? It seems unlikely. It's been around too long to just die out. It will just need to readjust, its purpose redefined. The writer and technologist Kevin Kelly once claimed that 'species of technology' are immortal; even seemingly extinct technologies are kept alive somewhere – either co-opted into some other format, reinvented as toys or playthings, or kept alive by hobbyists and enthusiasts. As Kelly writes:

> With very few exceptions, technologies don't die. In this way they differ from biological species, which in the long-term inevitably do go extinct. Technologies are idea-based, and culture is their memory. They can be resurrected if forgotten, and can be recorded (by increasingly better means) so that they won't be overlooked. Technologies are forever.

The invention of the light bulb meant that people stopped using candles to light their homes, but the candle didn't die – its purpose simply changed. It moved from technology to art and we see it now as romantic rather than a gloomy fire hazard. The crackly imperfection of vinyl became the warmth and charm of the object when compared to the CD or MP3. Consider the difference between the physical experience of holding a book in your hands, a bundle of paper and ink and glue, compared to its ebook equivalent (hello if you're reading this on Kindle by the way, you don't know what you're missing). The limitations of stationery – the fact that ink can smudge or that a page from a notebook can tear – are also part of its appeal. Unlike a file on a computer which can be endlessly duplicated and shared with the click of a button, a handwritten letter is a unique and personal object. Even just writing a phone number down on a Post-it Note requires some physical connection. The physical means something. People like it.

Even when moving to the digital world, people feel reassured by the physical. Skeuomorphic design – the replication of an object's physical characteristics in another material or form – has long been used by software designers so that users will immediately understand how to interact with new interfaces. Visual metaphors, such as the magnifying glass meaning 'search' or the nuts and bolts meaning 'settings', are easy to understand. They make sense because they relate to our real-world experiences. In *How We Became Posthuman*, N. Katherine Hayles described skeuomorphs as 'threshold devices, smoothing the transition between one conceptual constellation and another'. The language of the 'desktop' and its replication of a traditional office space on to a computer screen is a classic example. The concept was introduced by Apple with its Lisa computer system in 1983. Previewing the new computer system for *Byte* magazine before its launch, Gregg Williams quoted a computer engineer who observed that 'the computer has to do word processing, filing, electronic mail, everything'. Previously, the production, distribution and storage of documents had all involved separate processes, each with their own infrastructure (the typewriter, the orange FOR INTERNAL USE ONLY envelope, the filing cabinet), but now one small grey box could do it all.

Williams describes the value of the desktop metaphor, with its use of 'recognisable objects such as folders and reports' to reassure users that their data is safe:

> 'After all,' it seems to be telling you, 'computer files can mysteriously disappear, but folders, reports, and tools do not. If a file disappears, there's a logical explanation – either you threw it away or you filed it elsewhere. In either case, the situation is still under your control.'

Well, usually under your control, anyway.

Beyond the desktop metaphor, stationery has done pretty well out of skeuomorphic design: the paperclips used to attach documents to emails; the envelopes used to show new messages;

the pens, brushes, pencils and erasers used in Photoshop; the push-pins used to represent posts in Wordpress; the pen meaning 'compose new email'; the clipboard and scissors to cut and paste; note-taking apps designed like yellow legal pads; highlighters and sticky notes. The list goes on.

And of course, it's not just when attaching a document to an email that we have experienced a digital paperclip. Clippy, the animated character who infuriated millions of people by popping up to say 'It looks like you're writing a letter' as they were writing a letter in Microsoft Office made his first appearance in Microsoft Office 97 and was finally discontinued as part of the release of Office 2007. There were several Office Assistant characters (including a butler, a robot and a wizard), but Clippy was the default character and the one who provoked the strongest reaction in users. The character was designed by an illustrator called Kevan Atteberry who lives in Bellevue, Washington. Originally starting with around 260 characters designed by twenty artists, the character list for Office Assistant was extensively audience tested, until it was eventually whittled down to a group of ten. Of the ten, Atteberry had designed two, with Clippy the overall favourite.

Clippy is clearly based on a Gem design, but the proportions have been slightly altered to make room for the eyes (the two ends where the wire terminates are quite short compared with a regular Gem). Atteberry used this design as he considered it to be 'the most iconic paperclip', and by using the Gem, he not only reflected its iconic status, but strengthened it. The character design didn't allow much opportunity to give Clippy personality, and so the eyes and eyebrows were used to express emotion ('very powerful elements for conveying feelings', explains Atteberry). Initially, the fame of Clippy passed Atteberry by, not least because he's actually a Mac owner. However, as he visited clients and friends and saw them using Word, he realised how well known Clippy was. Well known, but not well loved. 'People either love him or hate him,' says Atteberry, 'it is very black and white. Though when the haters find out I designed him, they are apologetic that they hate him – but they still hate him.'

These visual metaphors can even offer a kind of digital afterlife to otherwise redundant practices, allowing anachronistic forms to live on – the language of 'cut and paste' developed by Larry Tesler and his team at the Xerox Corporation Palo Alto Research Center in the 1970s remains with us, although the physical process of cutting a paragraph of text from one page and pasting it on to another has faded from our office lives. Similarly, the 'call' feature on smartphones is represented by an old-style telephone receiver.

Under the leadership of Steve Jobs, Apple's design relied heavily on skeuomorphic design (the leather stitching of iCal was apparently based on the interior of his Gulfstream jet). However, with Jonathan Ive replacing Steve Forstall as head of Apple's Human Interface team, the reliance on real-world design elements was scaled back for the release of iOS 7 in 2013. Microsoft's Metro design language, used for Windows 8 and Windows Phone, appears to have been a deliberate attempt to differentiate itself from the skeuomorphic design over-used by Apple, focusing instead on typography and clean, flat design which is 'authentically digital'.

With skeuomorphic elements gradually being replaced by simpler, flatter designs on our tablets and smartphones, what could actually result is a greater appreciation of the physical (the real physical, not just the digital with a leather texture applied over the top). Without the crutch of skeuomorphs, the difference between writing in a notebook and writing on a tablet will become more pronounced. Both have their merits, but they are different activities done for different reasons. With one no longer adopting the drag of the other, both can flourish.

And so people who rush to announce the death of handwriting or those tech-evangelists looking forward to the singularity, the moment when artificial intelligence outsmarts human intelligence, should not get too excited. Stationery is not about to die. It's been around since the dawn of civilisation and it's not going to let some plucky upstart like the internet kill it off without a fight. And besides, a pen doesn't suddenly stop working just because you've gone into a tunnel; no one has ever needed to borrow a charger because the battery on their pencil has died; and if you're writing in a Moleskine, you never need to worry about having a bad signal or it crashing before you've had a chance to save your work.

The pen is not dead. Long live the pen.

Index

SELECT BIBLIOGRAPHY

Frank W. Abagnale, *The Art of the Steal: How to Protect Yourself and Your Business from Fraud* (New York: Broadway, 2001).
Nicholson Baker, *The Mezzanine: A Novel* (New York: Weidenfeld & Nicolson, 1988).
Elfreda Buckland and Donald McGill, *The World of Donald McGill* (London: Blandford, 1990).
Bruce Chatwin, *The Songlines* (New York: Viking, 1987).
Steven Connor, *Paraphernalia: The Curious Lives of Magical Things* (London: Profile, 2011).
Floyd L. Darrow, *The Story of an Ancient Art, from the Earliest Adhesives to Vegetable Glue* (Lansdale, PA: Perkins Glue, 1930).
David Diringer, *The Book before Printing: Ancient, Medieval, and Oriental* (New York: Dover Publications, 1982).
Giorgio Dragoni and Giuseppe Fichera (eds.), contributions by Giovanna D'Amia, Giorgio Dragoni, Giuseppe Fichera, Alessandra Ferretti, Hazel Juvenal-Smith, Armando Petrucci, Augusto Piccinini and Anna Ronchi, *Fountain Pens: History and Design* (Woodbridge: Antique Collectors' Club, 1998).
Henry Gostony and Stuart L. Schneider, *The Incredible Ball Point Pen: A Comprehensive History & Price Guide* (Atglen, PA: Schiffer Pub., 1998).
Philip Hensher, *The Missing Ink: The Lost Art of Handwriting (and Why It Still Matters)* (London: Macmillan, 2012).
Richard Herring and George Croly, *Paper & Paper Making, Ancient and Modern* (London: Longman, Brown, Green, and Longmans, 1855).
Richard Leslie Hills, *Papermaking in Britain, 1488–1988: A Short History* (London: Athlone, 1988).
John Wilfrid Hinde, introduction by Martin Parr, *Our True Intent Is All for Your Delight: The John Hinde Butlin's Photographs: Photography by Elmar Ludwig, Edmund Nägele and David Noble* (London: Chris Boot, 2002).
L. Graham Hogg, *The Biro Ballpoint Pen* (Southport: LGH Publications, 2007)
Virginia Huck, *Brand of the Tartan: the 3M Story* (New York: Appleton-Century-Crofts, 1955).
Dard Hunter, *Papermaking: The History and Technique of an Ancient Craft* (New York: Dover Publications, 1978).
György Moldova, trans. David Robert Evans, *Ballpoint: A Tale of Genius and Grit, Perilous Times, and the Invention That Changed the Way We Write* (North Adams, MA: New Europe, 2012).
Sonja Neef, *Imprint and Trace: Handwriting in the Age of Technology* (London: Reaktion, 2011).
Henry Petroski, *The Evolution of Useful Things: How Everyday Artefacts – from Forks and Pins to Paperclips and Zippers – Came to be as They are* (New York: Knopf, 1992).
Henry Petroski, *Invention by Design: How Engineers Get from Thought to Thing* (Cambridge, MA: Harvard UP, 1996).
Henry Petroski, *The Pencil: A History of Design and Circumstance* (New York: Knopf, 1990).
Herbert Spencer, *An Autobiography* (London: Williams & Norgate, 1904).
Frank Staff, *The Picture Postcard & Its Origins* (New York: F.A. Praeger, 1966).
Ethlie Ann Vare and Greg Ptacek, *Mothers of Invention: From the Bra to the Bomb: Forgotten Women & Their Unforgettable Ideas* (New York: Morrow, 1988).
Robert Wallsten and Elaine Steinbeck (eds.), *John Steinbeck: A Life in Letters* (Harmondsworth: Penguin, 1976).
Ian Whitelaw, *A Measure of All Things: The Story of Man and Measurement* (New York: St. Martin's, 2007).
JoAnne Yates, *Control through Communication: The Rise of System in American Management* (Baltimore: Johns Hopkins UP, 1989).